## Lecture Notes in Biomathematics

Vol. 1: P. Waltman, Deterministic Threshold Models in the Theory of Epidemics. V, 101 pages. 1974. DM 16,-

Vol. 2: Mathematical Problems in Biology, Victoria Conference 1973. Edited by P. van den Driessche. VI, 280 pages. 1974. DM 28,-

Vol. 3: D. Ludwig, Stochastic Population Theories. VI, 108 pages. 1974. DM 18,-

Lecture Notes in Biomathematics
Managing Editor: S. Levin

3

Donald Ludwig

# Stochastic Population Theories

Notes by Michael Levandowsky

Springer-Verlag
Berlin · Heidelberg · New York 1974

Dr. Donald Ludwig
Courant Institute of Mathematical Sciences
New York University
251 Mercer Street
New York, NY 10012/USA

**Present Address:**
University of British Columbia
Department of Mathematics
Vancouver, B.C., VGT 1W5
Canada

**Library of Congress Cataloging in Publication Data**

Ludwig, Donald, 1933-
Stochastic population theories.

(Lecture notes in biomathematics ; 3)
Bibliography: p.
Includes index.
1. Population biology--Mathematical models.
I. Title. II. Series.
QH352.L8 574.5'24'0184 74-23598

AMS Subject Classifications (1970): 92-02, 92A10, 92A15

ISBN-13: 978-3-540-07010-8 e-ISBN-13: 978-3-642-80883-8
DOI: 10.1007/ 978-3-642-80883-8

## Preface

These notes serve as an introduction to stochastic theories which are useful in population biology; they are based on a course given at the Courant Institute, New York, in the Spring of 1974. In order to make the material accessible to a wide audience, it is assumed that the reader has only a slight acquaintance with probability theory and differential equations. The more sophisticated topics, such as the qualitative behavior of nonlinear models, are approached through a succession of simpler problems. Emphasis is placed upon intuitive interpretations, rather than upon formal proofs. In most cases, the reader is referred elsewhere for a rigorous development. On the other hand, an attempt has been made to treat simple, useful models in some detail. Thus these notes complement the existing mathematical literature, and there appears to be little duplication of existing works.

The authors are indebted to Miss Jeanette Figueroa for her beautiful and speedy typing of this work.

---

The research was supported by the National Science Foundation under Grant No. GP-32996X3.

## CONTENTS

# I. LINEAR MODELS

General References: Bharoucha-Reid (1960), Feller (1951, 1966), Karlin (1969).

## I.1 The Poisson Process

Consider the number of fish caught in an interval of time (0,t) from a pond containing N fish. Assuming that N is large and t is small, and that each catch is an independent random event with probability rt, it follows that

$$P_0 = \text{Prob}\{\text{no fish are caught}\} = (1-rt)^N ,$$

(1)
$$P_1 = \text{Prob}\{\text{1 fish is caught}\} = Nrt(1-rt)^{N-1} ,$$

$$P_2 = \binom{N}{2}(rt)^2(1-rt)^{N-2} ,$$

$$P_i = \binom{N}{i}(rt)^i(1-rt)^{N-i} .$$

The generating function of the sequence is defined by

(2) $$F_N(t,x) = \sum_{n=0}^{N} P_n x^n .$$

From the form of the coefficients (1), it is clear that

(3) $$F_N(t,x) = (1-rt+rtx)^N .$$

Setting x = 1, we verify that $\sum_{n=0}^{N} P_n = 1$. The expected number of fish caught is

(4) $$\frac{\partial F}{\partial x}\Big|_{x=1} = \sum_{0}^{N} nP_n = Nrt .$$

If we assume that Nr = λ, where λ is independent of N, then

(5) $$F_N = \left(1 + \frac{\lambda t(x-1)}{N}\right)^N \xrightarrow[N\to\infty]{} e^{\lambda t(x-1)} .$$

Thus in the case of an infinite fish population our model becomes

(6) $$P_0 = e^{-\lambda t},\ P_1 = e^{-\lambda t}\lambda t,\ \ldots,\ P_r = e^{-\lambda t}\frac{(\lambda t)^r}{r!}$$

(7) $$F(t,x) = e^{\lambda t(x-1)}$$

(8) $$\frac{\partial F}{\partial x}\Big|_{x=1} = \lambda t = \sum_{n=1}^{\infty} nP_n ,$$

(9) $$\frac{\partial^2 F}{\partial x^2}\Big| = \lambda^2 t^2 = \sum n(n-1)P_n = \sum n^2 P_n - \sum nP_n$$

$$= \sum n^2 P_n - \lambda t ,$$

whence the variance is given by

(10) $$\sigma^2 = \sum n^2 P_n - (\sum nP_n)^2 = \lambda t .$$

So far we have considered the problem from the point of view of an observer watching a very large number of fishermen, possibly fishing in a very large number of ponds. Now, turning to the problems of the fisherman, we calculate the distribution of waiting times between catches. Thus if $C(t)$ is the number of fish caught by a fisherman up to time $t$, we shall calculate the distribution of jump points in a graph of this sort:

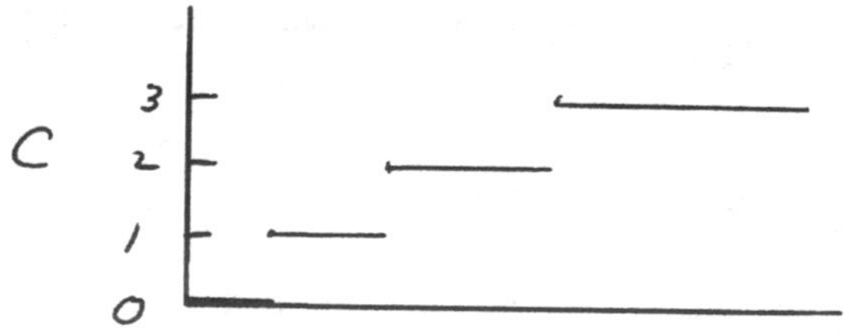


If the number of fish caught is a small fraction of the total number of fish, it is reasonable to assume that the behavior of the process is independent of its past history. That is,

$$\text{Prob[catching } (n_1+n_2) \text{ fish in time } (t_1+t_2)\ |n_1 \text{ were caught by time } t_1]$$

$$= \text{Prob[catching } n_2 \text{ in time } t_2] .$$

As before,

(12) $$P_0(t_2) = \text{Prob[no fish caught in interval } (t_1,t_1+t_2)] = e^{-\lambda t_2},$$

and for small $\delta t$,

(13) $$\text{Prob[catching a fish in the interval } (t_1+t_2,t_1+t_2+\delta t)] = \lambda\delta t + o(\delta t) .$$

Putting these together we obtain

(14) $$\text{Prob(waiting time lies in the interval } (t_1+t_2,t_1+t_2+\delta t)) = e^{-\lambda t_2}\lambda\delta t + o(\delta t)$$

That is, if T is the waiting time, in the limit

(15) $$\text{Prob}(t \leq T < t+dt) = \lambda e^{-\lambda t}dt ,$$

(16) $$E[T] = \int_0^\infty \lambda t e^{-\lambda t}dt = \frac{1}{\lambda} .$$

In a sequence of catches, let $T_j$ be the j-th waiting time. Then

(17) $$P_n(t) = \text{Prob}(T_1 + \ldots + T_n \leq t, \text{ and } T_1 + \ldots + T_{n+1} > t) .$$

This relationship is exploited in Karlin (1969), Ch. 9.

We now turn to the problem of computing $P_n(t)$ from their infinitesimal increments. Using the power series expansion of the exponential function, from (6), it follows that

(18) $$\text{Prob[no fish caught in } (t,t+\delta t)] = e^{-\lambda\delta t} = 1-\lambda\delta t + O(\delta t^2) ,$$

(19) $$\text{Prob[1 fish caught in } (t,t+\delta t)] = \lambda\delta t e^{-\lambda\delta t} = \lambda\delta t + O(\delta t^2) ,$$

(20) $$\text{Prob[2 or more fish caught in } (t,t+\delta t)] = O(\delta t^2) .$$

From this we see that

$$P_n(t+\delta t) = P_n(t)(1-\lambda\delta t + 0(\delta t^2)) + P_{n-1}(t)(\lambda\delta t + 0(\delta t^2)) + 0(\delta t^2) . \tag{21}$$

Rearranging terms, dividing by $\delta t$, and taking the limit as $\delta t \to 0$, this yields the system of ordinary differential equations

$$\frac{dP_n(t)}{dt} = -\lambda P_n(t) + \lambda P_{n-1}(t) , \quad n = 0 , \quad \ldots, \tag{22}$$

with initial conditions

$$P_0(0) = 1, \; P_j(0) = 0 \text{ for } j > 0 . \tag{23}$$

We can also derive the generating function $F(x,t)$ from this system. Summing the relations (22) after multiplication by $x^n$,

$$\sum x^n \frac{dP_n(t)}{dt} = \sum -\lambda x^n P_n(t) + \sum \lambda x^n P_{n-1}(t) , \tag{24}$$

which is equivalent to

$$\frac{\partial F(x,t)}{\partial t} = -\lambda F + \lambda x F = \lambda(x-1)F \tag{25}$$

with $F(0,x) = 1$.

This is readily integrated, and we see that

$$F(t,x) = e^{\lambda(x-1)t} , \tag{26}$$

in agreement with (5).

The Poisson process described here may be taken as a simple model of the number of encounters between any predator and its prey, or of the number of mutations appearing in a breeding population. It can also be used for spatial processes, such as the appearance of a plant species along a sampling transect. Such applications are discussed in the book by Karlin (Chapter 12).

In fact, observations of spatial distributions or of predator-prey encounters often show a deviation from this model (E. C. Pielou, 1969), but that does not necessarily diminish its usefulness. In such cases it can be viewed as a standard against which to test for non-random effects.

Problem:

Use the generating function to see when the Poisson distribution goes to the normal distribution. Other problems are given in Karlin, pp. 208-217.

## 2. Birth and Death Processes

### I.2.1 Linear Birth Process

References: A. G. McKendrick (1914), G. U. Yule (1924).

Now we apply the ideas of the previous section to the problem of estimating the growth of a large population of n organisms (for purposes of this discussion we consider only the fertile females), where all are assumed equally likely to reproduce. If population size at time t is n, then in analogy with the previous treatment we assume that

$$\text{(1)}\qquad \begin{aligned} &\text{Prob}[C(t+\delta t) = n + 1] = n\lambda\delta t + O(\delta t^2) , \\ &\text{Prob}[C(t+\delta t) = n] = 1-n\lambda\delta t + O(\delta t^2) , \\ &\text{Prob}[C(t+\delta t) > n + 1] = O(\delta t^2) . \end{aligned}$$

It follows that

$$\text{(2)}\qquad P_n(t+\delta t) = P_n(t)(1 - n\lambda\delta t + O(\delta t^2)) + P_{n-1}(t)(n - 1)\lambda\delta t + O(\delta t^2)) ,$$

which leads to the system of ordinary differential equations:

$$\text{(3)}\qquad \frac{d}{dt} P_n = - n\lambda P_n + (n - 1)\lambda P_{n-1} , \; n = 0, 1, \ldots$$

$$\text{(4)}\qquad P_1(0) = 1, \; P_j(0) = 0 \text{ if } j > 1 .$$

As before we calculate $F(t,x) = \sum P_n(t)x^n$ from a first order partial differential equation: by summation of (3) we have

$$\text{(5)}\qquad \sum x^n \frac{d}{dt} P_n = \sum - n\lambda x^n P_n + \sum (n-1)\lambda x^n P_{n-1} ,$$

or

$$\text{(6)}\qquad \frac{\partial}{\partial t} F = - \lambda x \frac{\partial F}{\partial x} + \lambda x^2 \frac{\partial F}{\partial x} ,$$

or

(7) $$\frac{\partial}{\partial t} F + x(1-x) \frac{\partial F}{\partial x} = 0$$

This is a first order partial differential equation of a particular simple kind.

In order to see this, we introduce the family of curves $x(s,x_o)$, $t(s,x_o)$ defined by

(8) $$\frac{dt}{ds} = 1 , \quad \frac{dx}{ds} = \lambda x(1-x) ,$$

(9) $$t(0,x_o) = 0 , \quad x(0,x_o) = x_o .$$

then (7) implies the ordinary differential equation

(10) $$\frac{dF}{ds} = \frac{\partial F}{\partial t} \frac{dt}{ds} + \frac{\partial F}{\partial x} \frac{dx}{ds} = 0 ,$$

with initial data $F(0,x)=x$. Of course, (8) is just the logistic equation. In order to solve it, let

(11) $$\xi = \log \left(\frac{x}{1-x}\right)$$

then, from (8),

(12) $$\frac{dx}{x(1-x)} = d\xi = \lambda ds .$$

Hence

(13) $$\xi = \xi_o + \lambda s ,$$

or

(14) $$\frac{x}{1-x} = \frac{x_o}{1-x_o} e^{\lambda s} = \frac{x_o}{1-x_o} e^{\lambda t} .$$

For given $x_o$, this defines a curve on which $F(t,x) = x_o$ is constant, since $\frac{dF}{ds} = 0$.

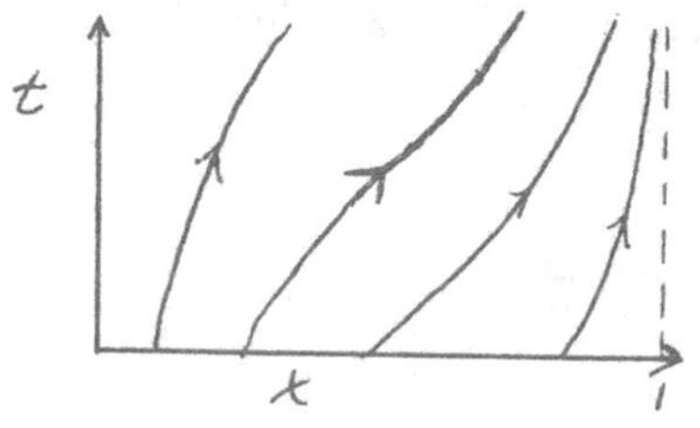

Solving (14) for $x_o$, we have

(15) $$F = x_o = \frac{xe^{-\lambda t}}{1 - x(1-e^{-\lambda t})}$$

Let

(16) $$y = x(1-e^{-\lambda t}) .$$

Then

(17) $$F = xe^{-\lambda t}\frac{1}{1-y} = xe^{-\lambda t}\sum_{j=o}^{\infty} y^j = \sum_{n=1}^{\infty} e^{-\lambda t}(1-e^{-\lambda t})^{n-1} x^n .$$

The coefficient of $x^n$ in (17) is just $P_n(t)$. Therefore

(18) $$P_n(t) = e^{-\lambda t}(1-e^{-\lambda t})^{n-1} .$$

In order to calculate the mean of the distribution, observe that

(19) $$\begin{aligned}\frac{d}{dt}\sum nP_n &= -\sum n^2\lambda P_n + \sum n(n-1)\lambda P_{n-1} \\ &= -\sum n^2\lambda P_n + \sum (n+1)n\lambda P_n \\ &= \lambda\sum nP_n\end{aligned}$$

Whence

(20) $$m(t) = \sum nP_n(t) = e^{\lambda t} .$$

Problem: Calculate the variance of this distribution.

Yule (1924) proposed the linear birth process as a model for the formation of new species within a given genus, and also for the formation of new genera within a family of such genera. There is an account of this work in Harris (1963), pp. 105-106.

## I.2.2 Linear Birth and Death Process

Now we extend the previous model to include death. As before, let C(t) be the number of fertile females. (We ignore variations of fertility with age, and the effects of population structure generally.) If C(t) = n, then the transition probabilities for a general birth and death process are given by

$$\text{Prob[birth in } (t,t+\delta t)] = \lambda_n \delta t + o(\delta t)$$

(1) $$\text{Prob[death in } (t,t+\delta t)] = \mu_n \delta t + o(\delta t)$$

$$\text{Prob[no change in } (t,t+\delta t)] = 1 - (\mu_n+\lambda_n)\delta t + o(\delta t)$$

where $\lambda_n$, $\mu_n$ are arbitrary.

Let $P_n(t) = \text{Prob}[C(t) = n]$. Then

(2) $$P_n(t+\delta t) = P_n(t)[1-\lambda_n\delta t-\mu_n\delta t] + P_{n-1}(t)[\lambda_{n-1}\delta t] + P_{n+1}(t)[\mu_n\delta t] + o(\delta t) ,$$

and, letting $\delta t \to o$,

(3) $$\frac{dP_n}{dt} = -(\lambda_n+\mu_n)P_n + \lambda_{n-1}P_{n-1} + \mu_{n+1}P_{n+1} .$$

Taking the simplest case first, let us assume that $\lambda_n$ and $\mu_n$ are linear in n, i.e. $\lambda_n = n\lambda$, $\mu_n = n\mu$. This corresponds to assuming that the probability of an individual giving birth or dying is not affected by the other members of the population. This is a good approximation for some populations, such as bacteria in the log-phase of growth.

The purpose of studying simple models of this sort is usually not to obtain detailed quantitative predictions, but rather to see the qualitative behavior of different sorts of model. For instance, in what follows we shall see qualitative differences in the behavior of this model and the corresponding deterministic one.

The generating function is given by

(4) $$F(t,x) = \sum P_n(t)x^n ;$$

it follows from (3) that

(5) $$\frac{dF}{dt} = -\sum P_n(t)x^n n(\lambda+\mu) + \sum P_{n-1}(n-1)x^n\lambda + \sum P_{n+1}(n+1)x^n\mu ,$$

and hence

(6) $$\frac{dF}{dt} = -x\frac{\partial F}{\partial x}(\lambda+\mu) + x^2\lambda\frac{\partial F}{\partial x} + \mu\frac{\partial F}{\partial x} ,$$

or

(7) $$\frac{\partial F}{\partial t} + ((\lambda+\mu)x - \lambda x^2 - \mu)\frac{\partial F}{\partial x} = 0 .$$

As before, we seek characteristic curves for (7) by setting

(8) $$\frac{dt}{ds} = 1 , \quad \frac{dx}{ds} = (\lambda+\mu)x-\lambda x^2-\mu = -\lambda(x-1)(x-\frac{\mu}{\lambda}) .$$

It follows that $\frac{dF}{ds} = 0$, i.e. F = constant on a characteristic curve.

If we set $\rho = \frac{\mu}{\lambda} = \frac{\text{death rate}}{\text{birth rate}}$, then (8) assumes the form

(9) $$\frac{dx}{(1-x)(x-\rho)} = \lambda ds .$$

If

(10) $$\xi = \log\frac{x-\rho}{1-x} ,$$

then

(11) $$d\xi = (\frac{1}{x-\rho} + \frac{1}{1-x})dx = \frac{1-\rho}{(1-x)(x-\rho)}dx = (1-\rho)\lambda ds = (\lambda-\mu)ds .$$

Let initial conditions for (8) be given as $x = x_0$, $t = 0$ at $s = 0$. Then $t = s$, and, from (11),

(12) $$\xi = \xi_0 + (\lambda-\mu)t ,$$

(13) $$\log\frac{x-\rho}{1-x} = \log\frac{x_0-\rho}{1-x_0} + (\lambda-\mu)t .$$

Along the characteristic curve, $F(t,x) = F(0,x_0)$, which can be evaluated in terms of the initial conditions for the process. If there is one individual at $t = 0$, then

(14) $$P_1(0) = 1, \; P_j(0) = 0 \text{ for } j \neq 1 ,$$

and hence $F(0,x_0) = x_0$.

Letting $\tau = (\lambda-\mu)t$ and solving for $x_0$, (13) and (14) lead to

(15) $$F(t,x) = \frac{\rho(1-x) + (x-\rho)e^{-\tau}}{1 - x + (x-\rho)e^{-\tau}} .$$

We see immediately from (15) that if $x = 1$, then $F = 1$

if $x = \rho$, then $F = \rho$

if $x \neq 1$, then $F \to \rho$ if $\tau \to +\infty$.

Therefore if $x \neq 1$ and $\lambda > \mu$, then $F \to \rho$ as $t \to \infty$. In view of (4), under the same assumptions we must have $P_0(t) \to \rho$, $P_j(t) \to 0$ for $j \neq 0$. Thus we see, for $\lambda > \mu$, the probability of extinction approaches $\rho = \frac{\mu}{\lambda}$ as $t \to \infty$. In the same way we see that for $\mu > \lambda$, $\tau \to -\infty$ as $t \to \infty$, and $P_0(t)$, the extinction probability, approaches 1, as we would expect.

Note: If we had initial conditions $C(0) = a$, then $P_j(0) = 0$ for $j \neq a$, and $P_a(0) = 1$. Hence, from (4), $F(o,x) = x^a$. The corresponding expression for $F(t,x)$ would be the one in (15) raised to the power a. We would then have an extinction probability $\rho^a$ if $\rho < 1$, corresponding to the joint extinction of a number of independent populations with the same parameter $\rho$.

Exercise 1. Expand F in a series and show that $P_n$ can be represented as

$$P_n(t) = P_1(t)R(t)^{n-1}$$

and that

$$P_1(t) \to 0 \text{ as } t \to \infty ,$$

$$R(t) \to 1 \text{ as } t \to \infty .$$

Exercise 2. Compute the mean and variance of this distribution and show that its mean is $e^{(\lambda-\mu)t}$ (this is the growth rate in the corresponding deterministic model), and its variance is (Bharoucha-Reid (1960))

$$\frac{1+\rho}{1-\rho} e^{(\lambda-\mu)t}(e^{(\lambda-\mu)t}-1) .$$

Sample Paths.

In our model, the population size $C(t)$ takes on a discrete set of values. We shall refer to a birth or death as a jump. Then

$$\text{Prob[jump in } (t,t+\delta t)] = (\lambda_n+\mu_n)\delta t + o(\delta t)$$

$$\text{Prob[no jump in } (t,t+\delta t)] = 1 - (\lambda_n+\mu_n)\delta t + o(\delta t) .$$

Then we know that the number of jumps obeys a Poisson process, and the waiting time between jumps has exponential distribution with

$$(16) \qquad \bar{T}_n = \text{mean waiting time} = \frac{1}{\lambda_n+\mu_n} .$$

Further, if a jump occurs, the probability of a birth is $\frac{\lambda_n}{\lambda_n+{}_n}$. Then, ignoring time for the moment, we can think of the process as a random walk, which can be represented graphically:

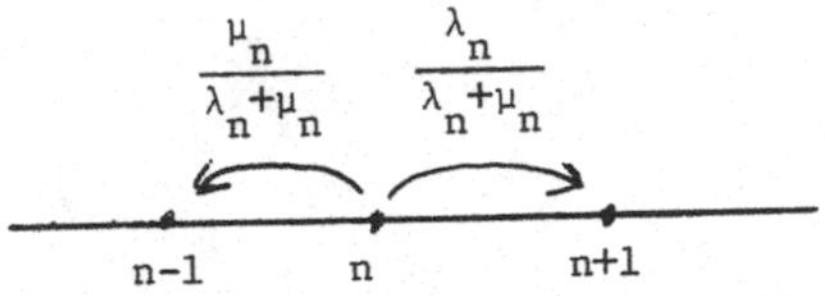


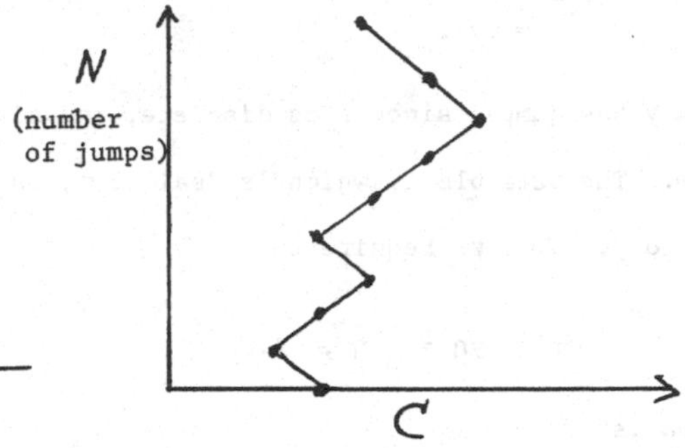


In the linear case the transition probabilities become simply $\frac{\lambda}{\lambda+\mu}$, $\frac{\mu}{\lambda+\mu}$. Then we can calculate probabilities for various events. Thus for a population of size 1, extinction in exactly 1 jump has probability $\frac{\mu}{\lambda+\mu} = \frac{\rho}{1+\rho}$; extinction in exactly 2 steps, $\frac{\lambda}{\lambda+\mu}\left(\frac{\mu}{\lambda+\mu}\right)^2$; etc. From such calculations we can find the probability of extinction: if $\rho \geq 1$ the probabilities of all such paths sum to 1, and if $\rho < 1$ they sum to $\rho$. One can prove this combinatorially, or simply appeal to the result of the previous section.

A Limit Theorem.

We turn now to the question of the fate of the proportion $1-\rho$ of populations (paths) which do not go to extinction. Our intention is to prove a limit theorem about these. Before proving this limit theorem we do a practice problem (assigned earlier as an exercise): Show that the Poisson distribution approaches a normal distribution as $\lambda t \to \infty$.

We begin with the Poisson generating function (see (1.26))

$$(17) \qquad F(t,x) = e^{\lambda t(x-1)} = \sum P_n(t)x^n ,$$

with mean $m = \sigma^2 = \lambda t$.

Let $x = e^{-\xi}$. The moment-generating function of the distribution is given by

(18) $$\Phi(t,\xi) = \sum_n P_n(t)e^{-n\xi} ,$$

i.e. the Laplace transform of the distribution. This may also be interpreted as the expectation of the random variable $e^{-\xi C}$, since $C(t)$ takes on the value n with probability $P_n(t)$. Now we introduce a new variable y such that

(19) $$n = my .$$

Although y has jumps, since n is discrete, the size of these jumps approaches zero as $m \to \infty$. The variable $\xi$, which is dual to n, is replaced by a variable $\eta$, which is dual to y. Thus we require that

(20) $$n\xi = y\eta = \frac{n}{m}\eta , \quad \text{i.e.}$$

$\eta$ must satisfy

(21) $$\xi = \frac{\eta}{m} .$$

Now we introduce the moment-generating function for the y-variable. This is most easily done by taking the expectation of $e^{-\eta Y}$, where Y is the random variable $C(t)/m(t)$. Thus we set

$$\Psi(t,\eta) = E(e^{-\eta Y}) .$$

From the definitions of $\eta$, Y and $\Phi$, we conclude that

$$E(e^{-\eta Y}) = E(e^{-\xi C}) = \Phi(t,\xi) = F(t,x) .$$

In view of the definitions of $\xi$ and $\eta$, we have

$$x = e^{-\xi} = e^{-\eta/m} .$$

Therefore (17) implies that

(22) $$\Psi(t,\eta) = \exp[\lambda t(e^{-\eta/m}-1)] = \exp[m(e^{-\eta/m}-1)] .$$

I.2.2 Linear Birth and Death Process

If $\eta$ is fixed and $m \to \infty$, then

$$e^{-\eta/m}-1 = -\eta/m + \frac{1}{2}\frac{\eta^2}{m^2} + \dots . \tag{23}$$

Therefore

$$\Psi(t,\eta) = \exp[-\eta + \frac{1}{2}\frac{\eta^2}{m} + \dots] \tag{24}$$

Thus, as $m \to \infty$,

$$\Psi \to \exp[-\eta + \frac{1}{2}\eta^2/m] . \tag{25}$$

Exercise 3. Check that this is the moment generating function of the normal distribution $N[1, \sigma^2 = (\lambda t)^{-1}]$.

Now we shall prove our limit theorem for the linear birth and death process. This calculation is good preparation for later proofs of more general theorems in the theory of branching processes. The calculation will be done in several steps, in imitation of the previous argument:

1. Introduce moment-generating function.
2. Scale n by the mean of the distribution.
3. Rescale the moment-generating function.
4. Pass to the limit.
5. Identify the result (invert Laplace transform).

Now we follow this programme for the conditional distribution of those populations that do not become extinct.

1. Moment generating function.

Returning to an earlier result (13), and recalling that for the probability generating function $F(x,t) = x_0$, we have

$$\log \frac{x-\rho}{1-x} = \log \frac{F-\rho}{1-F} + (\lambda-\mu)t . \tag{26}$$

In order to obtain a moment generating function, we let $x = e^{-\xi}$, $F(x) = \Phi(\xi)$, as in the practice problem just done.

2. Scaling by the mean.

The mean over all possible paths is $e^{(\lambda-\mu)t}$. However we want to use the mean over paths that don't lead to extinction. Since, for large t, the probability of extinction is nearly $\rho$, we may write

$$m(t) = 0 \cdot \rho + (1-\rho)m^*(t) , \tag{27}$$

where $m^*$ is the mean of the conditional distribution of those populations which don't become extinct. Thus $m^* = m/(1-\rho)$. Now we set

$$n = m^* y \tag{28}$$

$$\eta = m^* \xi , \tag{29}$$

in order to make

$$\eta\xi = y\eta . \tag{30}$$

3. Rescale moment generating function.

For this we rewrite (26), as in the practice problem:

$$\log \frac{e^{-\eta/m^*}-\rho}{1-e^{-\eta/m^*}} = \log \frac{(\Psi-\rho)}{(1-\Psi)} + (\lambda-\mu)t . \tag{31}$$

4. Pass to limit.

By taking the leading terms in the numerator and denominator of (31), we obtain

$$\log(1-\rho + \ldots) - \log(\eta/m^* \ldots) = \log \frac{\Psi-\rho}{1-\Psi} + (\lambda-\mu)t . \tag{32}$$

Since $\log m^* = (\lambda-\mu)t - \log(1-\rho)$, (32) becomes, after substituting and cancelling,

$$\log \frac{1}{(\Psi-\rho)} + \log \left(\frac{1-\Psi}{\eta}\right) = 0 , \tag{33}$$

or

$$\eta(\Psi-\rho) = 1 - \Psi , \tag{34}$$

whence

$$\Psi = \frac{1+\rho\eta}{1+\eta} , \tag{35}$$

$$\Psi - \rho = \frac{1-\rho\eta-\rho(1+\eta)}{1+\eta} = \frac{1-\rho}{1+\eta} . \tag{36}$$

5. Recognize the result.

We seek f(y) such that

$$\Psi - \rho = \int_0^\infty f(y)e^{-y\eta}dy \ . \tag{37}$$

A table of Laplace transforms, (or inspection) reveals that

$$\frac{1}{1+\eta} = \int_0^\infty e^{-y}e^{-y\eta}dy \ . \tag{38}$$

Comparing this result with (36), we see that

$$\Psi = \rho + (1-\rho)\int_0^\infty e^{-y}e^{-y\eta}dy \ . \tag{39}$$

This formula must represent the Laplace transform of the limit distribution. The term $\rho$ is the Laplace transform of a $\delta$-function concentrated at the origin. Thus formula (39) has the following interpretation: a proportion $\rho$ of the original ensemble of populations become extinct. The remaining proportion $1-\rho$ of the populations have a limiting exponential distribution, after scaling the size by $m^*(t)$.

<u>Exercise 4</u>. Verify this result directly by expanding $P_1(t)$, $R(t)$ (see exercise 1).

Comparing these results with the deterministic result, where $n(t) = e^{(\lambda-\mu)t}$, we find these qualitative differences:

(1) The deterministic result fails for the proportion $\rho$ of populations which become extinct.

(2) For the proportion $1-\rho$ that don't become extinct it gives the wrong value for the mean, namely $m$ instead of $m^*$.

(3) Though the correct order of magnitude of population size (overall mean) is given in the limit, the variance of the scaled variable y does not go to zero as $t \to \infty$. In this sense then the deterministic result is not the limit of the stochastic theory. According to exercise 2, the variance of the distribution of

population sizes is given by

$$\sigma^2 = \frac{1+\rho}{1-\rho} m(m-1)$$

Scaling by $m^*$, the variance of the scaled variable is

$$\sigma^{*2} = \frac{1+\rho}{1-\rho} \frac{m(m-1)}{m^2} (1-\rho^2) \to (1+\rho)(1-\rho)$$

Thus in the limit the stochastic result doesn't peak sharply, but remains "smeared" with a non zero variance.

### I.2.3 Birth and Death with Carrying Capacity

Reference: R. H. Mac Arthur and E. O. Wilson (1967).

The following theory arises naturally in the context of community ecology. One might attempt to predict the diversity of species on islands from data such as the island's size and distance from the mainland, and its variety of habitat. The word "island" here may refer to any sort of isolated or patchy habitat, such as mountain-tops or ocean deeps, as well as to the usual sort of island. In a typical observation, a naturalist might make a complete census of the bird species on an island on two occasions, ten years apart. Usually, some species will have disappeared and others appeared in the interval, but the total number often stays more or less constant. Evidently the islands are continually being colonized, and older lines are dropping out. A further test of theory is found in field experiments such as those of Simberloff and Wilson (1970), who defaunated entire mangrove islets and observed the rates of recolonization by insects and other organisms.

Mac Arthur applied a modified form of linear birth and death model to predict the number of individuals of a given species on an island. Assuming as a first approximation that an island has a constant carrying capacity K, we set the birth rate equal to zero when population size exceeds K. Otherwise the linear model is unchanged. Our main interest is in the qualitative properties of this admittedly crude model. Because of the nonlinearity of Mac Arthur's model for populations larger than K, there are really only K+2 possible states (including 0), and since from each of these there is a non-zero probability of extinction in at most K+1 steps, eventual extinction is certain for such a population.

Expected time to extinction.

Let $T_j$ be the expected time to extinction of a population starting with j individuals. This birth and death model can be considered a random walk, with expected time between jumps given by (16) of the previous section, where n is the current population size. Then since, when a jump occurs,

$$\text{(1)} \qquad \text{prob[birth]} = \frac{\lambda}{\lambda+\mu}, \quad \text{prob[death]} = \frac{\mu}{\lambda+\mu},$$

we have

$$\text{(2)} \qquad T_j = \frac{\lambda}{\lambda+\mu} T_{j+1} + \frac{\mu}{\lambda+\mu} T_{j-1} + \frac{1}{j(\lambda+\mu)},$$

(the last term being the waiting time for the next transition). After summing (2) over j, the result is

$$\text{(3)} \qquad \sum_{j=1}^{K} T_j = \frac{\lambda}{\lambda+\mu}\sum_{j=2}^{K+1} T_j + \frac{\mu}{\lambda+\mu}\sum_{j=1}^{K-1} T_j + \frac{1}{\lambda+\mu}\sum_{j=1}^{K}\frac{1}{j}$$

$$= \sum_{1}^{K} T_j - \frac{\lambda}{\lambda+\mu} T_1 + \frac{\lambda}{\lambda+\mu} T_{K+1} - \frac{\mu}{\lambda+\mu} T_K + \frac{1}{\lambda+\mu}\sum_{1}^{K}\frac{1}{j}.$$

But

$$\text{(4)} \qquad T_{K+1} = T_K + \frac{1}{\mu(K+1)}.$$

Therefore (3) implies that

$$\text{(5)} \qquad \lambda T_1 = \lambda\left(T_K + \frac{1}{\mu(K+1)}\right) - \mu T_K + \sum_{1}^{K}\frac{1}{j},$$

and hence

$$\text{(6)} \qquad T_1 = \frac{\lambda-\mu}{\lambda} T_K + \frac{1}{\lambda}\sum_{1}^{K}\frac{1}{j} + \frac{1}{\mu(K+1)}.$$

The last two terms grow slowly, ~ log K, and so we have for large K,

$$\text{(7)} \qquad T_1 \sim (1-\rho) T_K.$$

This formula (7) and our limiting results for the linear birth and death process support the following

Claim: Colonization in this model can be divided into two stages.

1. The early stage in which probability of extinction is approximately $\rho$ i.e., a a proportion $\rho$ of the populations become extinct fairly quickly.

2. The later stage, which is reached by a proportion $1-\rho$ of the populations. The carrying capacity K is reached (the population is established), and expected extinction time is much longer.

In order to obtain more precise results, we shall need a formula for $T_1$. To obtain this, we use the methods in Karlin (1969), sec. 7.7. Let

(8) $$z_j = T_j - T_{j-1} .$$

Then (2) may be rewritten as

(9) $$T_j = \frac{\lambda}{\lambda+\mu}(z_{j+1} + T_j) + \frac{\mu}{\lambda+\mu}(T_j - z_j) + \frac{1}{j(\lambda+\mu)} ,$$

which yields

(10) $$\lambda z_{j+1} - \mu z_j + \frac{1}{j} = 0 ,$$

or the system

(11) $$\begin{cases} z_{j+1} = \rho z_j + \frac{1}{\lambda j} , \\ z_1 = T_1 \end{cases}$$

From this we see that

(12) $$z_2 = \rho T_1 - \frac{1}{\lambda} ,$$

and so forth, so that

(13) $$z_{K+1} = \rho^K T_1 - \frac{1}{\lambda} \sum_{j=1}^{K} \frac{1}{j} \rho^{K-1} .$$

But since

(14) $$z_{K+1} = T_{K+1} - T_K = \frac{1}{\mu(K+1)} ,$$

(13) yields

$$T_1 = \frac{1}{\lambda}\sum_{j=1}^{K}\frac{1}{j}\rho^{-j} + \frac{\rho^{-K}}{\mu(K+1)}, \tag{15}$$

In the same manner we could also obtain a formula for $T_K$, from (6).

We shall only consider the case where $\frac{\mu}{\lambda} < 1$, since otherwise extinction will obviously be rapid. Since the largest term in (15) is of order $\rho^{-K}$, we should expect that $T_1$ will be large when $\rho^{-K}$ is large. In order to obtain a more precise result, we shall find an asymptotic form for the sum in (15). Let $\rho = 1/x$, and

$$f(x) = \sum_{j=1}^{K}\frac{1}{j}x^j . \tag{16}$$

then

$$f'(x) = \sum_{j=0}^{K-1} x^j = \frac{1-x^K}{1-x} .$$

Letting

$$I = \sum_{j=1}^{K}\frac{1}{j}\rho^{-j} = \int_0^{\frac{1}{\rho}} \frac{1-y^K}{1-y}\,dy , \tag{18}$$

we observe that for large K the major contribution to I is at the end point $y = \frac{1}{\rho}$.

Thus

$$I \sim \int^{\frac{1}{\rho}} \frac{y^K}{y-1}\,dy \sim \int^{\frac{1}{\rho}} \frac{y^K}{\frac{1}{\rho}-1}\,dy = \frac{\rho^{-K}}{(K+1)(1-\rho)} . \tag{19}$$

Thus, from (15), (18) and (19), we obtain

$$T_1 \sim \left[\frac{1}{\lambda(1-\rho)(K+1)} + \frac{1}{\mu(K+1)}\right]\rho^{-K} = \frac{1}{\mu(1-\rho)}\rho^{-K} . \tag{20}$$

This is better than the previous heuristic estimate, since we now have the coefficient of $\rho^{-K}$. The above can be rewritten as

$$T_1 \sim \frac{1}{\lambda(K+1)(1-\rho)}\rho^{-(K+1)} = \frac{1}{\lambda(K+1)(1-\rho)}\exp[(K+1)\log\frac{1}{\rho}] . \tag{21}$$

The exponent is the important part of (21); for $(K+1)\log\frac{1}{\rho} \sim 2, 3$ we will start to see large values of $T_1$. The single formula (21) summarizes the information contained in the figures on pp. 74-75 of MacArthur and Wilson (1967).

Since K, $\lambda$, $\mu$ and immigration rate are usually unknown, it is difficult to obtain a detailed test of the theory. However, the functional dependence on some parameters can often be observed; for example K or immigration rate may be proportional to an island's area in some cases. Immigration rate may also vary with distance from the mainland.

### I.3.1 Branching Processes (continuous time).

Reference: Th. E. Harris (1963).

Suppose that on dying an individual gives rise to J new individuals, where J is a random variable with

(1) $$h_j = \mathrm{Prob}[J = j] \ .$$

A well-known example in physics is the generation of neutrons by neutron-nucleus collision. Biological examples will be discussed later. We now examine population growth under this model.

As before, for population of size n(t),

(2) $$\mathrm{Prob}[\text{death in } (t,t+\delta t)] = n\beta\delta t + o(\delta t) \ ,$$

Let

(3) $$P_n(t) = \mathrm{Prob}[\text{population size is } n] \ .$$

Then

(4) $$P_n(t+\delta t) = P_n(t)(1-n\beta\delta t) + \sum_{j=o}^{\infty} (n+1-j)\beta\delta t h_j P_{n+1-j} + o(\delta t) \ ,$$

and hence

(5) $$\frac{dP_n}{dt} = - n\beta P_n + \sum_{j=o}^{\infty} (n+1-j)\beta \ h_j P_{n+1-j} \ .$$

We introduce the generating function

(6) $$F(t,x) = \sum P_n x^n \ .$$

The equation (5) can be combined as

(7) $$\sum x^n \frac{dP_n}{dt} = -\sum nx^n \beta P_n + \sum P_{n+1-j}(n+1-j)x^{n-j}\beta h_j x^j ,$$

which is equivalent to

(8) $$\frac{\partial F}{\partial t} = -\beta x \frac{\partial F}{\partial x} + \frac{\partial F}{\partial x} \sum_j h_j x^j .$$

Let us introduce the generating function for the distribution of offspring:

(9) $$h(x) = \sum h_j x^j .$$

Then (8) becomes

(10) $$\frac{\partial F}{\partial t} + \beta(x-h(x)) \frac{\partial F}{\partial x} = 0 .$$

We analyze this equation in the same way as equation (7) of I.2.2, by introducing a parameter s such that

(11) $$\frac{dt}{ds} = 1 , \quad \frac{dx}{ds} = \beta(x-h(x)) ,$$

with $\frac{dF}{ds} = 0$ along the curve parameterized by s. The solution can then be written in the form

(12) $$\int_{x_o}^{x} \frac{dy}{y-h(y)} = \beta t ,$$

which is a slight generalization of equations (9) and (13) of I.2.2 with a quadratic denominator.

As before,

(13) $$F(t,x) = F(0,x_o) .$$

If we start with a individuals at time t = 0, then

(14) $$F = x_o^a$$

The expected number of decendants of a single individual is given by

(15) $$\mu = h'(1) .$$

Exercise 1. Show that if x is near to 1, then (12) implies that

$$\frac{x_o - 1}{x - 1} \sim e^{\beta(\mu-1)t}$$

As a corollary to this we have

(16) $$F(t,1) \equiv 1 \; , \quad \frac{\partial F}{\partial x}(t,1) = ae^{\beta(\mu-1)t} = \sum nP_n(t) .$$

Exercise 2. Show that (a) if $\mu > 1$ and $h_o \neq 0$, then $h(x) = x$ at exactly one point $\rho$ in the open interval $(0,1)$. (b) if $\mu \leq 1$ then $h(x) = x$ has no roots in $(0,1)$. Of course, $h(1) = 1$.

Exercise 3. Show that if $0 \leq x < \rho$, then $x \leq x_o < \rho$, and if $\rho < x < 1$, then $\rho < x_o \leq x$. If $\mu > 1$, then $x_o \to \rho$ as $t \to \infty$. In fact, $x_o - \rho \sim C(x-\rho)e^{(1-\mu)\beta t}$ for large t.

Hint: Write $\frac{1}{y-h(y)} = \frac{A}{y-1} + \frac{B}{y-\rho}$ + regular function .

From exercise 2(b), we conclude that $F(t,x) \to 1$ as $t \to \infty$ if $\mu \leq 1$, i.e. eventual extinction is certain in that case. On the other hand, if $\mu > 1$, then the probability of extinction is $\rho$, according to exercise 3.

Now we shall derive a limiting distribution for the case $\mu > 1$, in analogy with the procedure in I.2.2. In analogy with equation (27) of that section, we have

(17) $$m(t) = \rho \cdot 0 + (1-\rho)m^*(t) .$$

Here $m(t)$ denotes the expected number of individuals in the population at time t, and $m^*$ denotes the same quantity, for the conditional distribution of sizes for populations which are not extinct. When (16) and (17) are combined, the result is

(18) $$m^*(t) = \frac{1}{1-\rho} e^{(\mu-1)\beta t} ,$$

if the population is started with a single individual at $t = 0$. Now, scaling as in Section 2.2, let $F(x) = \Psi(\eta)$, where

(19) $$x = e^{-\xi} = \exp[-\eta/m^*] .$$

I.3.1 Branching Process (continuous time)

Then (12) becomes

$$\int_{\Psi}^{e^{-\eta/m^*}} \frac{dy}{y-h(y)} = \beta t . \tag{20}$$

Since

$$e^{-\eta/m^*} = 1 - \frac{\eta}{m^*(t)} + \ldots \tag{21}$$

we see that the upper limit of integration goes to one, and since the denominator of the integrand vanishes there, we should pay particular attention to the behavior of the integrand at y = 1.

Define the regular function B(y) by

$$\frac{1}{y-h(y)} = \frac{1}{(y-1)(1-h'(1))} + B(y) . \tag{22}$$

Then (20) can be written as

$$\beta t = \int_{\Psi}^{1-\eta/m^*} \frac{1}{(y-1)(1-h'(1))} + \int_{\Psi}^{1-\eta/m^*} B(y)dy \tag{23}$$

$$= \frac{1}{1-h'(1)} [\log \frac{\eta/m^*}{1-\Psi}] + \int_{\Psi}^{1-\eta/m^*} B(y)dy$$

$$\sim \frac{1}{1-h'(1)} \{\log \eta - \log m^* - \log(1-\Psi)\} + \int_{\Psi}^{1} Bdy .$$

But since

$$- \log m^* = - \log \frac{m(t)}{1-\rho} = \log(1-\rho) - \beta t(h'(1)-1) , \tag{24}$$

this yields the limit formula

$$\frac{1}{1-h'(1)} [\log \eta + \log(1-\eta) - \log(1-\Psi)] + \int_{\Psi}^{1} B(y)dy = 0 . \tag{25}$$

This is as far as we can go toward the answer in the general case. Harris (1963) gives a proof that this integral equation has a solution that is a moment-generating function. The main value of this is perhaps to convince us that there is a limit distribution and tell us in which scale to look for it. That is, in order to compute this limit we may solve the original equations (5) for $P_n(t)$ and look at the cumulative distribution function in $y = \frac{n}{m^*}$. In this scale a limit will be approached rapidly.

I.3.2 Galton-Watson process (Branching process with discrete time)

Reference: H. W. Watson and Francis Galton (1874).

This process was first proposed in connection with attempts to understand why family names among the English peers disappeared over a number of generations. The question was, whether being a peer tended to diminish one's fertility. Another interpretation, due to R. A. Fisher, is the problem of survival of a mutant gene appearing in a population in the absence of selective forces. Many other interpretations are possible.

Since only males transmit the family name, we shall count only the number of males in each generation. Let n be the generation number. Let J be a random variable, which gives the number of sons sired by a man,

(1) $$f_j = \text{Prob}[J = j] \ .$$

The corresponding generating function is

(2) $$f(x) = \sum f_j x^j \ .$$

Now suppose the population size $X(n) = k$. What is the distribution of $X(n+1)$? Let $j_1$ , ..., $j_k$ be the number of sons of the 1,...,k-th individuals, respectively, in the n-th generation. Then

(3) $$\text{Prob}[X(n+1) = j] = \sum f_{j_1} \cdots f_{j_k} \ ,$$

where the summation is taken over all indices such that $j_1 + \ldots + j_k = j$. The generating function is given by

I.3.2 Galton-Watson Process

(4) $$F(n,x) = \sum P_i(n)x^i ,$$

where

(5) $$P_i(n) = \text{Prob}[X(n) = i] .$$

Let $F_k(n,x)$ be the conditional generating function, given that $X(n-1) = k$. Then,

(6) $$F_k(n+1,x) = \sum_j \sum_{j_1+\ldots+j_k=j} f_{j_1}\ldots f_{j_k} x_j$$

$$= \sum_{j=0}^{\infty} \sum_{j_1+\ldots+j_k=j} f_{j_1} x^{j_1} \ldots f_{j_k} x^{j_k}$$

$$= \sum_{j_1} f_{j_1} x^{j_1} \ldots \sum_{j_k} f_{j_k} x^{j_k} = (f(x))^k$$

Thus we see that, in general,

(7) $$F(n+1,x) = \sum P_k(n)(f(x))^k$$

$$= F(n,f(x)) .$$

If $X(0) = 1$, then (7) implies that

(8) $$F(1,x) = f(x)$$

$$F(2,x) = f(f(x))$$

.

.

.etc.

Often, $f(x)$ will be the generating function of the Poisson distribution.

(9) $$f(x) = e^{\lambda(x-1)}$$

<u>Exercise 1.</u> (a) Compute $F(n,x)$ if

$$f(x) = 1 - \frac{b}{1-C} + \frac{bx}{1-Cx}$$

(b) How is this related to the linear birth and death process?

Note that, if $F(0,1) = 1$ and $f(1) = 1$ then

(10) $$F(n,1) = 1,$$

and

(11) $$F'(n+1,1) = F'(n,1)f'(1) .$$

Hence

(12) $$F'(n,1) = F'(0,1)(f'(1))^n .$$

Note that $f'(1) = \sum if_i$ is the expected number of sons. The mean population size grows geometrically. There are 3 cases:

(a) if $f'(x) > 1$ the expected population size grows

(b) if $f'(x) < 1$ the expected population size decreases

(c) if $f'(x) = 1$ the expected population size is constant.

Now we investigate extinction. If $X(0) = 1$ let $\rho$ = probability of eventual extinction. In order for the family line founded by an individual to become extinct, the subfamilies founded by each of his sons must become extinct. If the founder has j sons, the probability of this event is $\rho^j$. Therefore

(13) $$\rho = \sum f_j \rho^i ,$$

we see that $\rho$ is a solution of the equation $\rho = f(\rho)$. The extinction probability $\rho$ is therefore determined by the intersection(s) of the straight line $y = x$ and the convex function $y = f(x)$. These are two cases:

$f'(1) \leq 1$

$f'(1) > 1$

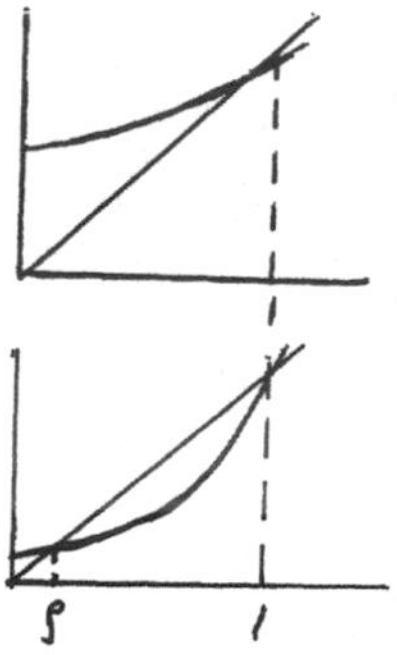


Thus if the mean is $\leq 1$ extinction is certain, as we would expect, but otherwise $\rho < 1$. We see also that for $x < \rho$, $\rho > f(x) > x$. Since

(14) $$F(1,x) = f(x) , \quad F(2,x) = f(f(x)) , \quad \text{etc.} ,$$

$F(n,x)$ in this case is monotone increasing. Similarly, if $x > \rho$, $F(n,x)$ is decreasing. Thus $F(n,x) \to \rho$ as $n \to \infty$. Looking at the convergence in more detail,

I.3.2 Galton-Watson Process

let $x_o = 0$. Then

(15) $$x_1 = f(0) = f_o = \text{probability of extinction by generation 1.}$$

(16) $$x_2 = f(f_o) = \sum_j f_j f_o^j = \text{probability of extinction by generation 2,}$$

and in general

(17) $$x_n = f(x_{n-1}) = \text{probability of extinction by generation n.}$$

This is true for all values of the mean, f'(1).

If f'(1) > 1, the limiting distribution of population sizes may be obtained by methods analogous to Section 3.1. For the problem of survival of family names (or mutant genes) the case f'(1) = 1 is of greatest interest. Therefore the following discussion is confined to that case. In this case, we have seen that $x_n \to 1$ as $n \to \infty$. Extinction is certain, but how fast does it occur?

Let $y_n = 1 - x_n$. Then $y_n \to 0$ as $n \to \infty$. Equation (17) may be rewritten as

(18) $$1 - y_{n+1} = f(1-y_n) = 1 - f'(1)y_n + \frac{f''(1)}{2} y_n^2 + \dots ,$$

Let

(19) $$\beta = \frac{f''(1)}{2} ,$$

and recall that f'(1) = 1. Then (18) becomes

(20) $$y_{n+1} = y_n - \beta y_n^2 , \text{ approximately.}$$

This quadratic difference equation is reminiscent of the Ricatti differential equation. That observation motivates the following analysis: Let

(21) $$z_n = \frac{1}{y_n}$$

Then (20) becomes

(22) $$\frac{1}{z_{n+1}} = \frac{1}{z_n}(1 - \frac{\beta}{z_n}) ,$$

or

(23) $$z_{n+1} = \frac{z_n}{1-\beta/z_n} = z_n(1 + \frac{\beta}{z_n} + \dots) = z_n + \beta , \text{ approximately .}$$

Thus, $z_n \sim n\beta$, or

(24) $$x_n \sim 1 - \frac{1}{n\beta} = \text{probability of extinction after n generations.}$$

Now let $m^*(n)$ be the mean of the conditional distribution of populations not extinct after n generations. Then, in analogy with equation (27) of Section 2.2, we have

(25) $$1 = x_n \cdot 0 + (1-x_n)m^* ,$$

or

(26) $$m^* = n\beta .$$

Thus the mean of the conditional distribution goes to infinity as $n \to \infty$. This explains how families can become extinct with probability one, but there are still plenty of people alive. One actually sees in some countries that a few names are extremely common, and similar phenomena may be seen in the growth of cities and corporations.

Next, we seek a limiting distribution for families that are not extinct. The approach is the same as before. We study the generating function through the relation (18):

(27) $$F(n+1,x) = f(F(n,x)) .$$

Let

(28) $$x = e^{-\xi} ; \quad \xi = \eta/m^* .$$

Then

(29) $$x = e^{-\eta/m^*(n)} \sim 1 - \frac{\eta}{m^*(n)}$$

We shall expand f about 1, since x is near 1. Let

(30) $$g_n = 1 - F(n,x) .$$

Then we expect $g_n$ to become small as $n \to \infty$, and so we insert (30) into (27) and expand f, to obtain

(31) $$1 - g_{n+1} = f(1-g_n) = 1 - f'(1)g_n + \frac{f''(1)}{2} g_n^2 + \dots ,$$

which we rewrite as

(32) $$g_{n+1} = g_n - \beta g_n^2 .$$

This equation is completely analogous to (20).

As before, we introduce the reciprocal

(33) $$h_n = \frac{1}{g_n} .$$

Then (32) becomes

(34) $$\frac{1}{h_{n+1}} = \frac{1}{h_n}(1 - \frac{\beta}{h_n})$$

$$h_{n+1} = \frac{h_n}{1-\beta/h_n} \sim h_n(1+\beta/h_n) = h_n + \beta .$$

This is valid for all n if x is close to 1. Now fix $\eta$. For large N

(35) $$f(0,x) = x \sim 1 - \frac{\eta}{m^*(N)} .$$

Then, from (34), we obtain

(36) $$h_n \sim h_0 + n\beta .$$

Furthermore,

(37) $$h_o = \frac{1}{g_o} = \frac{1}{1-x} \sim \frac{1}{1-(1-\eta/m^*(N))} = \frac{m^*(N)}{\eta} .$$

Putting (30), (33), (36) and (37) together, we have (for large N)

(38) $$F(N,x) = 1 - \frac{1}{h_N} = 1 - \frac{1}{\frac{m^*}{\eta} + m^*} = 1 - \frac{\eta}{(1+\eta)m^*} .$$

Now, in the limit we seek an expression of the type

(39) $$F = x_N + (1-x_N)Q(\eta) .$$

But we really have such an expression: from (24) and (26)

(40) $$x_N = 1 - \frac{1}{m^*(N)} .$$

Furthermore

(41) $$- \frac{\eta}{1+\eta} = - 1 + \frac{1}{1+\eta} .$$

If (41) is employed in (38), the result is

$$F(N,x) = 1 - \frac{1}{m^*(N)} + \frac{1}{m^*(N)} \cdot \frac{1}{1+\eta} . \tag{42}$$

Comparison of (42) with (40) shows that (39) is valid, with

$$Q(\eta) = \frac{1}{1+\eta} \int_0^\infty e^{-\eta y} e^{-y} dy . \tag{43}$$

Thus the limiting conditional distribution for families which do not become extinct is exponential.

<u>Exercise:</u> Derive the analogous formula for the case of continuous time, if $h'(1)=1$.

This concludes our treatment of linear stochastic models. It is worthwhile to pause here and re-examine their general relation to the deterministic models. They all correspond to exponential growth of the form

$$x = x_o e^{mt} ,$$

or

$$x_n = x_o m^n .$$

That is, in all cases we have

$$\sum jP_j(t) = e^{mt} .$$

This is a consequence of linearity, and is not true for more general stochastic models.

If $m < 1$, both deterministic and stochastic models predict extinction, so deterministic theory is to that extent adequate in this case.

If $m > 1$, as we saw in the case of linear birth and death, a proportion $\rho$ of the populations go to extinction. The remaining proportion $1-\rho$ grows exponentially, in the sense that after exponential scaling a limit distribution is approached. This distribution is known explicitly only for the case of linear birth and death, or linear fractional generating function. (In other cases it is easily computed however).

If $m = 1$, extinction is certain in the long run. However after scaling proportional to n there is (again) an exponential limit distribution, with a rather long tail. In genetics, this last observation leads one to expect a small proportion of neutral alleles to establish themselves. It would be interesting to know how this is related to the high genetic heterogeneity that is often seen in natural populations. (See Murray (1972)).

# II. EPIDEMICS

## II.1 The Reed-Frost Model

Reference: N. T. J. Bailey (1957).

This is the simplest nonlinear epidemic model. We assume there is an irreversible succession of classes to which an individual can belong:

| S | → | I | → | R |
|---|---|---|---|---|
| Susceptible | | Infective | | Removed |

Disease is spread by effective contact between single infective and susceptible individuals. Contacts are assumed to be independent random events, the probability of contact in a given interval being the same for all possible pairs of susceptible and infective individual (homogeneous mixing). The infectious period is fixed, and taken as the unit of time. Since

(1) $$S + I + R = N = \text{Total population size} ,$$

it suffices to know $S_n$, $I_n$ in order to specify the state of the population at time n. We define

(2) $$P_{S,I}(n) = \text{Prob}[S \text{ susceptibles, } I \text{ infectives at time } n] .$$

Consider a given susceptible. For each infective let the probability of making contact be p. Then the probability of no contacts is

(3) $$Q_n = (1-p)^{I_n} = q^{I_n} .$$

Thus the number of infectives at time n+1 has a binomial distribution:

(4) $$P_{S_n,0}(n+1) = Q^{S_n} .$$

(5) $$P_{S_{n-1},1}(n+1) = S_n Q_n^{S_{n-1}} (1-Q_n)$$

and

(6) $$P_{S_n-I_{n+1},I_{n+1}}(n+1) = \binom{S_n}{I_{n+1}} Q_n^{S_n-I_{n+1}} (1-Q_n)^{I_{n+1}} ,$$

We see that if $I_n$, $S_n$ and p are known we can compute

(7) $$E(I_{n+1}) = S_n(1-Q_n) ,$$

(8) $$E(S_{n+1}) = S_n Q_n .$$

II.1.1 Deterministic Version of the Reed-Frost Model

The previous model suggests an analogous deterministic theory

(9) $$i_{n+1} = s_n(1-q^{i_n})$$

(10) $$s_{n+1} = s_n q^{i_n}$$

where $i_o$, $s_o$ are given.

Exercise Show that, in general, $i_n \neq E(I_n)$, $s_n \neq E(S_n)$. That is, the analogous deterministic model doesn't yield the expected values of the stochastic model.

We examine some results of the deterministic model:

$$s_1 = s_o q^{i_o}$$

(11) $$s_2 = s_1 q^{i_1} = s_o q^{i_o+i_1}$$

$$\vdots$$

$$s_{n+1} = s_o q^{i_o+i_1+\dots+i_n}$$

Let $N \to \infty$, and $i_n \to 0$ in (11). Thus we obtain

(12) $$s_\infty = s_o q^{\sum_o^\infty i_k} = s_o q^{N-s_\infty} .$$

In order to see the limiting properties of $s_\infty$, we use scaled variables. Let

(13) $$\sigma_\infty \sim \frac{s_\infty}{N} , \quad \sigma_o = \frac{s_o}{N} , \quad q^N = e^{-k} .$$

Then

$$(14) \qquad \sigma_\infty = \sigma_o q^{N(1-\sigma_\infty)} = \sigma_o e^{-(1-\sigma_\infty)k} .$$

In order to analyze the situation graphically, let us assume first that $\sigma_o$ is one. Then for various values of k we have two qualitatively different cases, which are familiar from the treatment of branching processes in I.3.

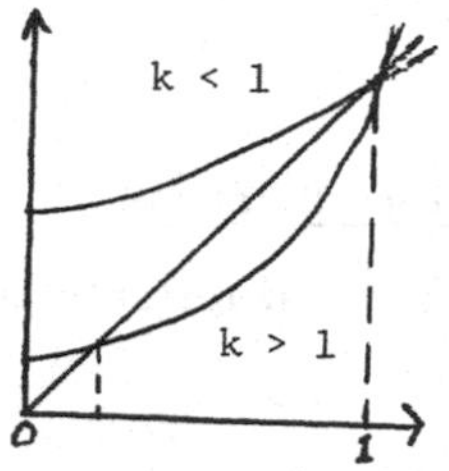


Graph of $f(\sigma) = e^{k(\sigma-1)}$ (the generating function of the Poisson distribution).

The quantity $\rho$ in the graph (page 34) solves $\rho = e^{k(\rho-1)}$ and is the final proportion of susceptibles. More realistically, we can now consider the effect of taking $_o$ slightly less than 1, graphically:

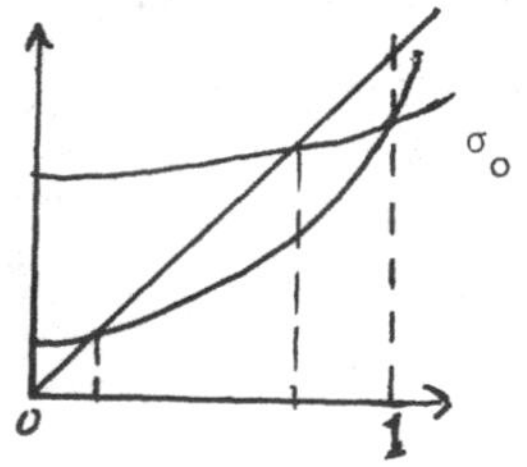


In this case the curves are slightly shifted down, and for given $k > 1$ there is another possible final proportion of susceptibles, $\sigma_\infty$, near 1. Thus we have a threshold theorem:

for $k < 1$ , $\sigma_o$ near 1 $\Longrightarrow$ $\sigma_\infty$ near 1

for $k > 1$ , $\sigma_o$ near 1 $\Longrightarrow$ $\sigma_\infty$ near $\rho$

(This is analogous to the threshold in the Kermack-McKendrick theory of epidemics. See Bailey (1957), Waltman (1974) or Hoppensteadt (1974)). The threshold value of k is therefore 1. Since

(15) $(1-p)^N = e^{-k}$

we have $1-p = e^{-k/N} \sim 1 - k/N$, i.e.

(16) $p \sim k/N$ .

Now, from the stochastic theory we have

(17) $Q_o = (1-p)^{I_o} = (1-k/N)^{I_o} = e^{I_o \log(1-k/N)} \sim e^{-kI_o/N} \sim 1 - \frac{kI_o}{N}$

Then

(18) $E(I_1) = S_o(1-Q_o) \sim S_o \frac{k}{N} I_o$

From this we see that the threshold parameter k is the expected number of effective contacts involving a given infective, and the threshold value is $k = 1$.

Summarizing from this deterministic model we can see that

1. $p \sim k/N$.
2. if $k < 1$ then $S_\infty \sim N$. No large epidemic occurs.
3. if $k > 1$ then $S_\infty \sim \rho N$. There will be a large epidemic, with a proportion $\rho$ of survivors.
4. $\rho \to 1$ as $k \to 1$.

Thus, we would predict that if $k < 1$ a population is safe, but if $k > 1$ it is vulnerable, and immunization campaigns or the like are indicated. The model is of course highly idealized in many respects. In particular it ignores population structure and the possibility of highly vulnerable pockets within a general population, which can serve as a source of recurrent epidemic outbreaks.

## II.1.2 Two Methods for the Study of the Reed-Frost Model

The main objective of our treatment of epidemics is the qualitative theory of II.2. Before proceeding to this rather lengthy discussion, we shall briefly examine some other methods for the study of stochastic epidemics.

Method I. Direct computation

For fixed $I_o$, $S_o$, p (setting k > 1 so that an epidemic occurs), and $n \leq S_o$, the numbers $P_{S,I}(n)$ are computed. This means that $O(N^3)$ numbers are computed, since $I + S \leq S_o$, $N \geq S_o$. This may be feasible for $N \leq 100$ or so.

Exercise 1. Give an algorithm for computing $P_{I,S}(n)$. When this calculation is done, we can compute the "expected epidemic" specified by

$$\bar{I}(n) = \sum_{I,S} P_{S,I}(n)I$$

$$\bar{S}(n) = \sum_{I,S} P_{S,I}(n)S$$

Such calculations show quite clearly that $\bar{I}$ and $\bar{S}$ are far from the deterministic values. Aside from that comparison, the results appear to have little real value, since there are not many data to which they can be compared. It is probably more useful to compute the distribution of final epidemic sizes; these correspond to the numbers $P_{S,o}(\infty)$ (= $P_{S,o}(N+1)$ for given N). The results give answers to questions such as

(a) What is the probability of a large epidemic, for given $I_o$, $S_o$, p?

(b) If there is a large epidemic, how large is it likely to be?

For such questions, however, data usually refer to a limited number of trials, so testing the theory is difficult. A disadvantage of method I is that there are 3 parameters to vary, and it is easy to bury oneself in computer output.

Method II. Monte-Carlo simulation

We approximate the solution by simulating the epidemic again and again, collecting statistics on the outcomes. First we pick $I_o$, $S_o$, p. Then the (binomial) distribution of $I_1$ is computed and used to divide the interval [0,1] into $S_o+1$ intervals. Picking a uniformly distributed random number R in [0,1], we choose $I_1$ to correspond to the interval in which R lies. Continuing in this way until the epidemic ends, we not only obtain statistics on final results, but can examine sample paths. Another advantage is that the process is easy to program and requires little storage space (O(N)).

Exercise 2. Give an algorithm for method II

Some further advantages of method II:

— One frequently gains insight by looking at sample paths generated in this way.

— It is much more flexible than method I. Introduction of complications such as a randomly varying infectious period is relatively easy here, but virtually impossible for method I.

— Surprisingly good accuracy can be obtained with a relatively small number of runs (e.g. 100 - 1000).

Some disadvantages:

— Greater accuracy is obtained very slowly with large numbers of runs. This is because error decreases as $M^{-1/2}$, where M is the number of runs. Because of these diminishing returns, 3 place accuracy may be prohibitively expensive.

— There is a certain ham-handedness in using such an empirical approach; when such calculations are performed to the exclusion of other methods, they seldom provide much insight into the problem.

## II.1.3 The Backward Equations

Reference: D. Ludwig (1974a)

Since the final size distribution is the main object of interest, we set out to find

$$P(R|S) = \text{Prob}[R \text{ removals starting with } S \text{ susceptibles}] ,$$

using the parameters $I_o$, p. The general procedure is most easily understood after examining some simple cases: S = 1.

$$(1) \qquad P(0|1) = Q_o = (1-p)^{I_o} = q^{I_o} ,$$

$$(2) \qquad P(1|1) = 1 - P(0|1) .$$

S = 2.

(3) $$P(0|2) = Q_o^2 ,$$

(4) $$P(1|2) = 2P(1|1)Q_o q ,$$

since $Q_o$ is the probability of avoiding the initial infectives and $q = 1-p$ is the probability of avoiding one infective. Then

(5) $$P(2|2) = 1 - P(0|2) - P(1|2) .$$

Similarly,

S = 3.

(6) $$P(0|3) = Q_o^3$$

(7) $$P(1|3) = 3P(1|1)Q_o^2 q^2$$

(8) $$P(2|3) = 3P(2|2)Q_o q^2$$

(9) $$P(3|3) = 1 - P(0|3) - P(1|3) - P(2|3) .$$

and for general S,

(10) $$P(0|S) = Q_o^S$$

(11) $$P(R|S) = \binom{S}{R} P(R|R) Q_o^{S-R} q^{R(S-R)} \quad , \quad (S < R) ,$$

(12) $$P(S|S) = 1 - \sum_{R=o}^{S-1} P(R|S) .$$

Advantages of Method III:

— Only $O(N^2)$ numbers are computed.

— The intermediate results are useful.

Disadvantages:

— The numbers $P(S|S)$ approach zero rapidly as $S \to \infty$ and so precision tends to be lost in the use of (12). Thus, with single precision on the CDC 6600 the method is only good to about $N = 50$.

But the main advantage is an unexpected benefit arising from study of the method, namely that it is immediately applicable to a wide range of models. Let the infectivity now be some more general function $\beta(t)$, so that

II.1.3 The Backward Equation

$$\text{Prob[Contact in } (t,t+\delta t)] = \beta(t)\delta t + o(\delta t) .$$

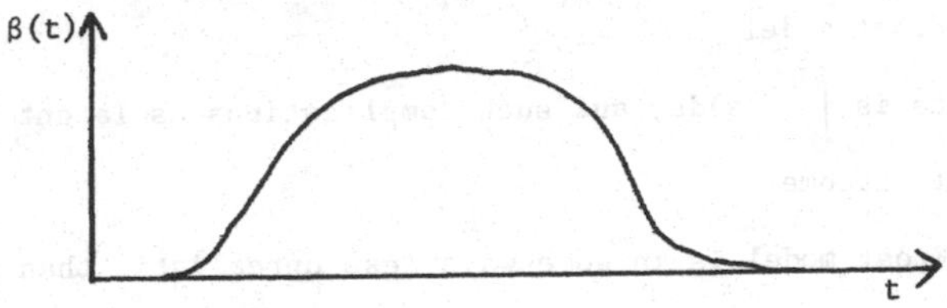


First we consider the simplest case, where I = 1 and S = 1.

Define

(13) $$C(t) = \text{Prob[No contact in } (0,t)] .$$

Then

(14) $$C(t+\delta t) = C(t)[1-\beta(t)\delta t] + \ldots ,$$

(15) $$\frac{d}{dt} C(t) = -\beta(t)C(t) ,$$

and hence

(16) $$C(t) = \exp(-\int_0^t \beta(s)ds) .$$

We set

(17) $$C(\infty) = q .$$

Now, using this q, we can follow precisely the same procedure as before, first for the case $I_o$ infectives, 1 susceptible.

(18) $$P(0|1) = q^{I_o} = Q_o$$

(19) $$P(1|1) = 1 - Q_o$$

Then for $I_o$ infectives, 2 susceptibles

(20) $$P(0|2) = Q_o^2$$

(21) $$P(1|2) = 2P(1|1)Q_o q ,$$

and in general

(22) $$P(R|S) = \binom{S}{R} P(R|R) Q_o^{S-R} q^{R(S-R)} ,$$

and so forth. From this we see that

1. The final size distribution for the more general epidemic is identical to that for the Reed-Frost model.

2. What counts is $\int_0^\infty \beta(s)ds$, and such complications as latent periods have no effect on the final outcome.

3. The Reed-Frost model is in some ways less unrealistic than first appears. (Actually, the simplest form of it may not be a bad detailed description of certain types of bacteriophage growth.)

This unexpected windfall from Method III means that, for testable results (the distribution of final sizes) it is not necessary to attempt the difficult task of specifying the population state in some general model which keeps track of the time since infection for each infective. Instead we merely use the results of the Reed-Frost model, which are the same. This shows the value of concentrating on "observable" (testable) variables. For generalized models,then, Method I is hopeless. Method II works if the model is discretized, and provides some insight. Method III is clearly best.

There is a generalization of the above results, which is given here without proof (see Ludwig (1974a)).

Theorem: To any epidemic model there corresponds a model that is discrete in time, with a finite number of states, which has the same final size distribution.

This theorem is useful for dealing with more than one type of infective. An example is polio, where most infections are "silent" (no symptoms, or mild symptoms that are ignored) and probably not important in speading the infection. We approach this problem as follows:

Let the type u be a random variable with density f(u)du, and $\beta(t,u)$ be the infectivity. Then the probability of no contact between a susceptible and infective of type u is given by

$$q(u) = e^{-\int_0^\infty \beta(t,u)dt} \quad . \tag{23}$$

II.1.3 The Backward Equation

Let

(24) $$Q_1 = E(q) = \int q(u)f(u)du$$
$$\vdots$$
$$Q_j = E(q^i)$$

Then for $I_o = 1$, $S = 1$, where the type of infective is unknown

(25) $$P(0|1) = Q_1 \ .$$

For $I_o = 1$, $S = 2$

(26) $$P(0,S) = E(q^2) = Q_2 \neq Q_1^2 \text{ , in general .}$$

For $I_o = 1$, general S,

(27) $$P(0,S) = E(q^S) = Q_S \ .$$

For $I_o > 1$, we proceed as before, starting with

(28) $$P(0|1) = Q_1^{I_o} \ .$$

The formulas become more complicated in this case. In this way the transition probabilities can be obtained, and then the final size distribution can be computed, e.g. by Method I. (See Ludwig (1974a)).

Another desirable generalization would be to have several types of susceptible. This occurs for example with flu, where some people are partially immune to some strains. We can deal with this in a similar manner.

Let the type V of susceptible be a random variable with density g(v)dv. Then we proceed as above, taking

$$\bar{q} = \int q(v)g(v)dv \text{ , etc.}$$

Details are given in Ludwig (1974a).

## II.2 Qualitative Theory for the "General Stochastic Epidemic" (Method IV)

References: P. Whittle (1955), D. G. Kendall (1956).

A drawback in Methods I-III is that they are useful only for small values of N, up to 100 or so. We now turn to a complementary method that is useful for large N. For convenience in the computations, we shall switch to a model with continuous time. This model has been called the "General Stochastic Epidemic" but that is poor terminology for it is actually rather special. It is analogous but not equivalent to the Reed-Frost model, and as before we have the sequence of transitions $S \longrightarrow I \longrightarrow R$. As before I = number of infectives, S = number of susceptibles, and R = number removed, with I + S + R = N (fixed).

Consider a pair consisting of 1 infective and 1 susceptible. Let

(1) $$\text{Prob}[\text{contact in } (t,t+\delta t)] = \beta\delta t + o(\delta t) ,$$

where by "contact" we mean as before, that the susceptible becomes infected. Let

(2) $$\text{Prob}[\text{infective} \longrightarrow \text{removed in } (t,t+\delta t)] = \gamma\delta t + o(\delta t) .$$

It should be noted that in the continuous model there are an infinite number of possible infective types, since the infective period has an exponential distribution. We define the state of the epidemic by the pair (I,R), but later it will be convenient to switch to (S,R). Then, setting $S_{I,R} = N - R - I$,

(3) $$P_{I,R}(t+\delta t) = P_{I,R}(t)(1-\beta I S_{I,R}\delta t-\gamma I\delta t) + P_{I-1,R}(t)\beta(I-1)S_{I-1,R}\delta t + P_{I+1,R-1}(t)\gamma(I+1)\delta t + o(\delta t) .$$

Now we scale the time, letting

(4) $$\beta' = N\beta , \quad t' = \beta' t , \quad \rho = \gamma/\beta'$$

Then, letting $\delta t \to o$, (3) becomes

(5) $$\frac{d}{dt'} P_{I,R} = -(I \frac{(N-I-R)}{N} + \rho I)P_{I,R} + (I-1)(\frac{N-I+1-R}{N})P_{I-1,R} + \rho(I+1)P_{I+1,R-1} .$$

The initial conditions are

$$(6)\qquad \begin{cases} P_{I,R}(0) = 0 \text{ if } I \neq 1 \text{ or } R \neq 0 , \\ P_{1,0}(0) = 1 . \end{cases}$$

2.1 Approximation of the epidemic by a birth and death process.

For large N, in the early states of an epidemic we shall have $I << N$, $R << N$, and it is reasonable to replace $\frac{N-I-R}{N}$ by 1. Then (5) is replaced by the simpler equation

$$(7)\qquad \frac{d}{dt'} P_{I,R} = -I(1+\rho)P_{I,R} + (I-1)P_{I-1,R} + \rho(I+1)P_{I+1,R-1} .$$

(we shall drop the prime and write t for t' from here on, for convenience). Now let

$$(8)\qquad P_I = \sum_R P_{I,R} .$$

Then, after summation over R, (7) becomes

$$(9)\qquad \frac{d}{dt} P_I = -I(1+\rho)P_I + (I-1)P_{I-1} + \rho(I+1)P_{I+1} .$$

Here we recognize the equations of the linear birth and death process, (equation (3) of I.2.2, or (5) of I.3.1), and therefore we can immediately write down some important facts about them. For instance, according to I.2.2, the extinction probability is obtained from the roots of

$$(10)\qquad h(x) = -(1+\rho)x + x^2 + \rho = (x-1)(x-\rho) .$$

Thus $\rho$ is the extinction probability. So, the results from the linear birth and death model would suggest that

(A) A proportion $\rho$ of the epidemics will terminate quickly and involve a small number of individuals.

(B) In a proportion $1-\rho$ of the epidemics, $I/m^*$ has an exponential limiting distribution, where

$$(11)\qquad m^*(t) = \frac{1}{1-\rho} e^{(1-\rho)t}$$

(Note: This is the same idea as in MacArthur's colonization model, and suggests a possible improvement on it. In the context of epidemics the idea is due to Whittle (1955) and Kendall (1956)).

Now we seek the limiting distribution via the generating function, as before. Let

(12) $$F(t,x,y) = \sum P_{I,R}(t)x^I y^R .$$

Then (7) implies that

(13) $$\frac{\partial F}{\partial t} = -(1+\rho)x\frac{\partial F}{\partial x} + x^2\frac{\partial F}{\partial x} + \rho y\frac{\partial F}{\partial x} .$$

The characteristic curves for (13) are given by

(14) $$\frac{dx}{dt} = -(x^2 - (1+\rho)x + \rho y) ,$$

(15) $$\frac{dy}{dt} = 0 , \quad \text{since } \frac{\partial F}{\partial y} \text{ is absent from (13) .}$$

The roots of the polynomial $\dot{x}$ are

(16) $$x_\pm = \frac{1+\rho}{2} \pm \frac{1}{2}\sqrt{(1+\rho)^2-4\rho y} .$$

In view of (6), $F(o,x,y) = x$, and hence, since $F$ is constant along the characteristics,

(17) $$F(t,x,y) = x_o .$$

In order to solve (14), we rewrite it in the form

(18) $$\frac{-dx}{x^2-(1+\rho)x+\rho y} = \frac{-dx}{(x_+-x_-)}\left(\frac{1}{x_+-x} + \frac{1}{x-x_-}\right) = dt .$$

This leads to a solution

(19) $$F = \frac{x_-(x_+-x) + x_+(x-x_-)\tau}{x_+-x + (x-x_-)\tau}$$

where

(20) $$\tau = e^{-(1-\rho)t} .$$

In order to find the limiting distribution we now let

(21) $$x = e^{-\xi/m^*}, \quad y = e^{-\eta/m^*}, \quad m^* = \frac{e^{(1-\rho)t}}{1-\rho} .$$

(In this scaling of y, we are following the reasonable assumption that the number of removals grows exponentially also). Then the result is

(22) $$F \sim \rho + \frac{1-\rho}{1+\xi+\frac{\rho}{1-\rho}\eta} .$$

<u>Exercise</u>. Prove this claim.

We now interpret the second term in (22) as a generating function.

(23) $$\frac{1}{1+\xi+\frac{\rho}{1-\rho}\eta} = \int_0^\infty e^{-u}e^{-u\xi-u\frac{\rho}{1-\rho}\eta}\,du = E[e^{-U\xi}e^{-U\frac{\rho}{1-\rho}\eta}] ,$$

if the random variable U has an exponential distribution. For the limiting distribution, the density of I, R is concentrated on the straight line

(24) $$R = \frac{\rho}{1-\rho} I ,$$

i.e. it is one-dimensional. Such results are typical for multi-type branching processes. The result (22), (23) corresponds to

(25) $$I \sim \frac{U}{1-\rho} e^{(1-\rho)t} , \quad R \sim U\frac{\rho}{(1-\rho)^2} e^{(1-\rho)t} .$$

Thus, in some sense,

(26) $$\frac{dR}{dt} \sim \rho I ,$$

which could have been foreseen from (2), after the scaling (4). The result (25) is similar to a deterministic result, corresponding to the birth and death model. In order to exploit this fact, we turn to the nonlinear deterministic theory.

### II.2.2 <u>Deterministic theory (Kermack and McKendrick)</u>

Let

(1) $$i = \frac{I}{N}, \quad r = \frac{R}{N}, \quad s = \frac{S}{N} .$$

Then

(2) $$i + s + r = 1 .$$

Assumptions (1) and (2) at the beginning of II.2 suggest that

(3) $$\frac{ds}{dt} = -is , \quad \frac{di}{dt} = is - \rho i , \quad \frac{dr}{dt} = \rho i$$

where $i \cdot s$ is the rate of infection, and $\rho i$ is the rate of removal. By dividing the first and third equations, we obtain

(4) $$\frac{ds}{dr} = - \frac{s}{\rho} ,$$

and therefore

(5) $$se^{r/\rho} \text{ is constant.}$$

If we set $s_\infty = s(\infty)$, then in the limit we have $i(\infty) = 0$, $r(\infty) = 1 - s_\infty$. From (5), it follows that

(6) $$s_\infty e^{\frac{1-s_\infty}{\rho}} = \text{constant} = s(0) ,$$

or

(7) $$s_\infty = s_0 e^{\frac{s_\infty - 1}{\rho}} .$$

Remarkably, this is the same result as for the Reed-Frost model. (See II.1.1). We can, again, look at this graphically. For $s_0 = 1$,

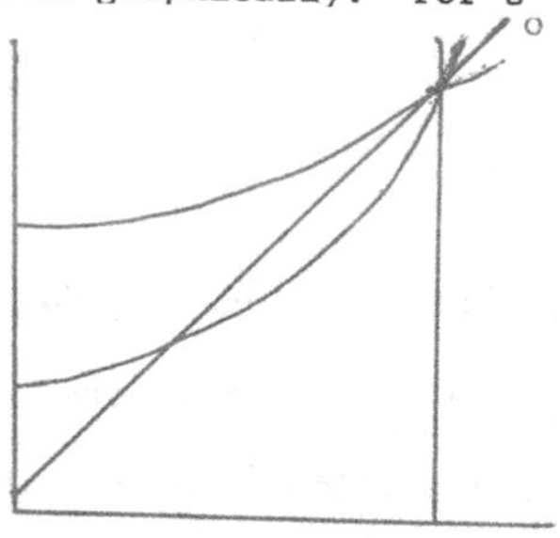

So we see that $\rho = 1$ is the threshold value, which of course we could have predicted. But $\rho$ has now two distinct roles: it is a threshold parameter, and it is the extinction probability for the approximating birth and death process. If $s_0 < 1$, slightly, the curves shift down slightly and $s_\infty$ moves slightly. That is, if

(8) $\Delta s_o \sim 0(\frac{1}{N})$ , then $\Delta s_\infty \sim 0(\frac{1}{N})$ .

Returning to the stochastic theory, we can now say that those epidemics which don't terminate quickly seem to grow more or less deterministically, and their final sizes vary only by $0(\frac{1}{N})$. Thus we can give answers to two of the important questions mentioned earlier:

1. What is the probability of a large epidemic?

   Answer: $1 - \rho$ .

2. If a large epidemic does occur, how large is it likely to be?

   Answer: $N(1 - s_\infty)$ .

We would expect a final size distribution of the kind shown by the unbroken line below.

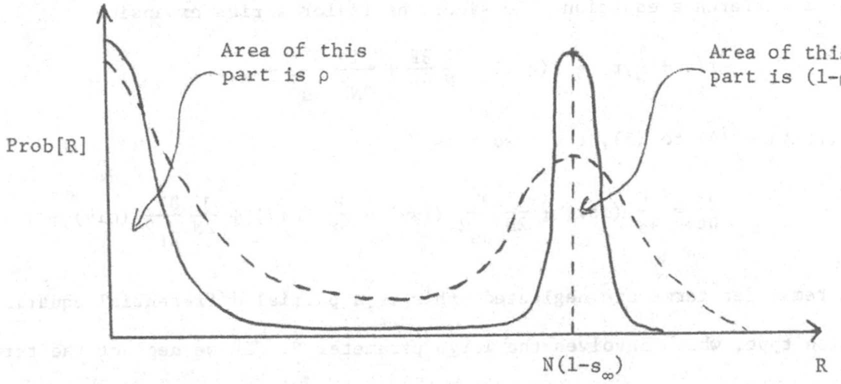


The broken curve superposed on the graph here represents the type of results given by exact (numerical) solutions. These give a wider spread about the point $N(1 - s_\infty)$ than the quasi-deterministic theory, of order $\sqrt{N}$. The general shapes of the two graphs are in agreement; a U-shaped distribution of final sizes is observed. The larger spread in the exact results suggests that a diffusion approximation may be an improvement over the quasi-deterministic theory.

## 2.3 Diffusion Approximation for the General Stochastic Epidemic

We shall rewrite the original equations ((5) of II.2) in terms of scaled variables. In terms of R, S, the equations are

(1) $\frac{d}{dt} P_{S,R} = -(N-S-R) \frac{S+\rho(N-S-R)}{NN} P_{S,R} + \frac{N-S+1-R}{N}(S+1)P_{S+1,R} + \rho(N-S-R+1)P_{S,R-1}$

Exercise. Derive these equations.

Let

(2) $$v(t,s,r) = P_{S,R}(t) ,$$

where

$$s = \frac{S}{N} , \quad r = \frac{R}{N} .$$

Then (1) may be rewritten in the form

(3) $$\frac{\partial}{\partial t} v(t,s,r) = -N(1-s-r)(s+\rho(1-s-r))v + N[(1-s-r)sv]\Big|_{s+\frac{1}{N},r} + N[\rho(1-s-r)v\Big|_{s,r-\frac{1}{N}} .$$

This is a difference equation. Consider the Taylor series expansion

(4) $$f(s + \frac{1}{N},r) = f(s,r) + \frac{1}{N}\frac{\partial f}{\partial s} + \frac{1}{2N^2}\frac{\partial^2 f}{\partial s^2} + \dots .$$

After applying (4) to (3), the result is

(5) $$\frac{\partial v}{\partial t} = \frac{\partial}{\partial s} (isv) + \frac{1}{2N}\frac{\partial^2}{\partial s^2} (isv) - \frac{\partial}{\partial r} (\rho iv) + \frac{1}{2N}\frac{\partial^2}{\partial r^2} (\rho iv) + \dots .$$

If the remainder terms are neglected, this is a partial differential equation of the diffusion type, which involves the large parameter N. If we neglect the terms divided by N, we obtain a first order partial differential equation

(6) $$\frac{\partial v}{\partial t} - \frac{\partial}{\partial s} (isv) + \frac{\partial}{\partial r} (\rho iv) = 0 .$$

The characteristics of (6) are

(7) $$\dot{t} = 1 , \quad \dot{s} = -is , \quad \dot{r} = \rho i .$$

These are the Kermack-McKendrick equations of II.2.2. In contrast to the previous cases which we have considered, $\dot{v} \neq 0$, since (6) can be written in the form

(8) $$\dot{v} + (\frac{\partial}{\partial s} (-is) + \frac{\partial}{\partial i} (\rho i))v = 0 .$$

Given initial values for t,s,r and v, equations (7) and (8) determine a unique trajectory in (t,s,r) space. In order to interpret v, let us consider the region D filled by the trajectories emanating from a closed bounded set $B_o$ in r,s plane (t = 0). The projection of this upon either the (t,0,r) or the (t,s,0) plane looks like

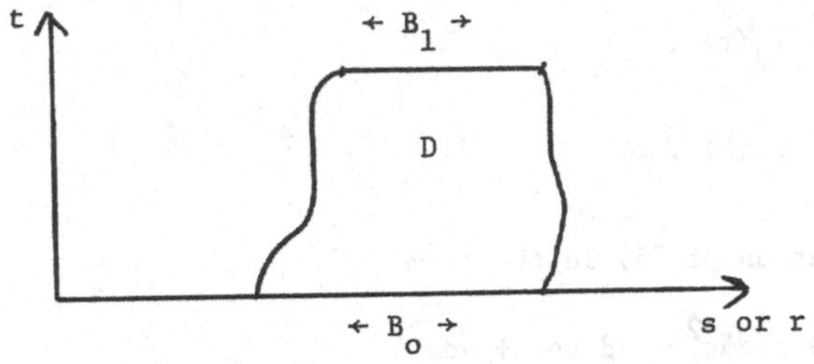


Let $B_1$ be the intersection of these trajectories with a plane $(t_1,s,r)$, ($B_1$ is a set of points at a later time $t_1$ which are joined to points of the initial set $B_o$ by trajectories.)

The region D is bounded by $B_o$, $B_1$, and the two dimensional surface formed by trajectories emanating from the boundary points $\partial B_o$ of $B_o$. The left-hand side of (8) is the divergence of the vector $(v,-isv,\rho iv)$. From Gauss' theorem (see for example, R. Courant (1936)).

(9) $$\iiint_D \nabla \cdot (v,-isv,\rho iv) = \iint_{\partial D} (v,-isv,\rho iv) \cdot dS$$

But on the part of $\partial D$ formed by trajectories from $\partial B_o$ the inner product in the right-hand integral is zero, and so we obtain the important result

(10) $$\iint_{B_o} v\,dr\,ds = \iint_{B_1} v\,dr\,ds .$$

The left-hand side of (10) is the probability of finding the population in $B_o$ at time 0, and the right-hand side is the probability of finding the population in $B_1$ at time $t_1$. A consequence of this is that, if v is concentrated at a point $s_o,r_o$ when t = 0 (i.e. if $B_o$ is shrunk to the point $s_o,r_o$), then at later time s is concentrated at points on the trajectory $s_d(t)$, $r_d(t)$ which emanates from $s_o,r_o$. Thus, according to (6), v is completely determined by the behavior of the trajectories (7).

Now we shall consider the modification of this picture brought about by the diffusion equation (5). If initially v is concentrated at $s_o, r_o$, we shall expect to see a Gaussian distribution centered at the point $s_d(t)$, $r_d(t)$ at later times. Let

(11) $$\delta s = s - s_d(t) ,$$

$$\delta r = r - r_d(t) .$$

Then we seek a solution of (5) in the form

(12) $$v \sim Ce^{-N/2(\alpha\delta r^2 + 2\beta\delta r\delta s + \gamma\delta s^2)}$$

where C is a normalization factor. Our task will be to determine the coefficients $\alpha, \beta, \gamma$. In order to illustrate the method of attack, we first do an easier practice problem.

4. Practice Problem

Consider the one-dimensional diffusion equation

(13) $$\partial_t v + \partial_x(b(x)v) = \frac{1}{2N}$$

The deterministic trajectory $x = x_d(t)$ satisfies

(14) $$\dot{t} = 1 , \quad \dot{x} = b(x) .$$

Now let $\delta x = x - x_d(t)$. Let us try a solution of (13) of the form suggested by (12):

(15) $$v = e^{-N\phi(x,t)}z(x,t) .$$

The factor $z(x,t)$ may be necessary in order to normalize the solution. Then, neglecting terms small compared to large N, we have

(16) $$\begin{cases} \partial_t v = - N\phi_t v + 0(1) , \\ \partial_x(bv) = - N\phi_x bv + 0(1) \\ \partial_x(av) = - N\phi_x av + 0(1) \\ \partial_x^2(av) = + N^2\phi_x a\phi_x v + 0(N) \end{cases}$$

II.3 Diffusion Approximation

Then by substituting (16) into the original equation (13), we see that $\phi$ must satisfy

(17) $$\phi_t + b\phi_x + \frac{1}{2} a\phi_x^2 = 0 .$$

This is a nonlinear partial differential equation of first order. It can be solved by means of the Hamilton-Jacobi theory. If we have an equation of the form (Hamilton-Jacobi equation)

(18) $$\phi_t + H(x,\phi_x) = 0 ,$$

then the corresponding system of Hamilton's equations is

(19) $$\dot{t} = 1 , \quad \dot{x} = \frac{\partial H}{\partial \phi_x} , \quad \dot{\phi}_x = - \frac{\partial H}{\partial x} .$$

In the present case this approach leads to the equations

(20) $$\begin{cases} \dot{t} = 1 \\ \dot{x} = b(x) + a(x)\phi(x) \\ \dot{\phi}_x = - b_x\phi_x - \frac{1}{2} a_x\phi_x^2 . \end{cases}$$

Note that if $\phi_x = 0$ initially, then $\phi_x = 0$ on the whole trajectory. It then follows that $x = b(x)$, i.e. Hamilton's equations (20) reduce to (14). We would in fact expect that $\phi_x = 0$ along the deterministic trajectory, if $\phi$ corresponds to a Gaussian centered on the trajectory. Since $\phi_x = 0$, it follows from (17) that also $\phi_t = 0$ on the deterministic trajectory, i.e. $\phi = 0$.

We should like to find the second derivative $\phi_{xx}$ on the deterministic path, since this would give us the variance. Now, letting $x = x_d + \delta x$, we consider the expansions

$$\phi(t,x) = \phi(t,x_d) + 0 + \frac{1}{2} \phi_{xx}(\delta x)^2 + \dots$$

$$b(x) = b(x_d) + b_x(x_d)\delta x + \dots$$

$$a(x) = a(x_d) + a_x(x_d)\delta x + \dots$$

$$\phi_x(x,t) = 0 + \phi_{xx}(x_d)\delta x + \frac{1}{2} \phi_{xxx}(\delta x)^2 + \dots$$

$$\phi_t(x,t) = 0 + \phi_{tx}\delta_x + \frac{1}{2} \phi_{txx} \frac{(\delta x)^2}{2} + \dots$$

Plugging these expressions into the original equation, we obtain

$$\phi_{tx}\delta x + \frac{1}{2}\phi_{txx}(\delta x)^2 + (b+b_x\delta x)(\phi_{xx}\delta x + \frac{1}{2}\phi_{xxx}(\delta x)^2) + \frac{1}{2}a(\phi_{xx}\delta x)^2 = 0 . \tag{21}$$

The linear terms in $\delta x$ yield

$$\phi_{xt}\delta x + b\phi_{xx}\delta x = 0 , \tag{22}$$

or

$$\dot{\phi}_x = 0 .$$

Therefore, the linear terms in (21) vanish, since $\phi_x = 0$ on the deterministic trajectory. After retaining only terms quadratic in $\delta x$ in (21) and then dividing by $\frac{1}{2}\delta x^2$, we obtain finally the equation

$$\phi_{txx} + b\phi_{xxx} + 2b_x\phi_{xx} + a(\phi_{xx})^2 = 0 . \tag{23}$$

The first two terms of this may be thought of as the derivative of $\phi_{xx}$ along the deterministic trajectory:

$$\dot{\phi}_{xx} = \phi_{xxt}\dot{t} + \phi_{xxx}\dot{x} .$$

Equation (23) therefore has the form of an ordinary differential equation along the deterministic trajectory:

$$\dot{\phi}_{xx} + 2b_x\phi_{xx} + a\phi_{xx}^2 = 0 . \tag{24}$$

This is a Ricatti equation, and as before we use the change of variable $\omega = \frac{1}{\phi_{xx}}$ to obtain the new equation

$$\dot{\omega} = 2b_x\omega + a \tag{25}$$

<u>Exercise 1</u>. Solve this equation, using an integrating factor

Now the solution (15) has the form

$$v(t,x) = \exp\left[\frac{-N(x-x_d)^2}{2\omega(t)}\right]z(t,x) . \tag{26}$$

Thus $\frac{\omega}{N}$ is the variance of the density v. If $b_x = 0$, then (25) reduces to

$$\dot{\omega} = a . \tag{27}$$

In this special case, the variance would be linearly increasing with t. This completes our analysis of the practice problem. We shall turn to the real problem in the next section. The method is exactly the same, but the notation becomes cumbersome because more variables are involved.

## 5. Gaussian approximation for a general diffusion equation

Because of the advantages afforded by a systematic use of indices it turns out that the calculations for the Gaussian approximation are easiest in the general case. Let $x = (x^1,\ldots,x^n)$. We adopt the tensor notation convention of summation over repeated indices. Then consider the equation

$$(1) \qquad v_t + \partial_j(b^j v) = \frac{1}{2N}\,\partial_j\partial_k(a^{jk}v)\ .$$

We try the solution

$$(2) \qquad v = e^{-N\phi(x,t)}z(x)\ .$$

This gives

$$(3) \qquad \begin{cases} v_t = -\,N\phi_t v + \ldots \\ \partial_j(b^j v) = -\,N\phi_j b^j v + \ldots \\ \partial_i(a^{ik}v) = -\,N\phi_i a^{ik} v + \ldots \\ \partial_j\partial_k(a^{jk}v) = +\,N^2\phi_j\phi_k a^{jk} v + \ldots \end{cases}$$

As before, (1) and (3) imply that

$$(4) \qquad \phi_t + b^j\phi_j + \frac{1}{2}\,a^{jk}\phi_j\phi_k = 0\ .$$

The deterministic trajectory $x_d(t)$ satisfies

$$(5) \qquad \frac{dx^i}{dt} = b^i(x)\ .$$

As before, we set

$$(6) \qquad \delta x^i = x^i - x_d^i(t)\ .$$

Expanding everything in sight,

$$
(7)\quad
\begin{cases}
\phi_j = \phi_{jk}\delta x^k + \frac{1}{2}\,\phi_{jk\ell}\delta x^k\delta x^\ell + \dots\,, \\
\phi_t = \phi_{tk}\delta x^k + \phi_{tk\ell}\dfrac{\delta x^k\delta x^\ell}{2} + \dots\,, \\
b^j = b^j(x_d) + b^j_k\delta x^k + \dots\,, \\
a = a(x_d) + \dots\,.
\end{cases}
$$

Substituting (7) into (4), the result is (up to quadratic terms)

$$
(8)\qquad \phi_{tk}\delta x^k + \phi_{tk\ell}\,\frac{\delta x^k\delta x^\ell}{2} + (b^j + b^j_\ell\delta x^\ell)(\phi_{jk}\delta x^k + \frac{1}{2}\,\phi_{ik\ell}\delta x^k\delta x^\ell) + \frac{1}{2}\,a^{jm}\phi_{j\ell}\delta x^\ell\phi_{km}\delta x^k + \dots = 0\,.
$$

Dealing first with linear terms we find that

$$
(9)\qquad \phi_{tk}\delta x^k + b^j\phi_{jk}\delta x^k = 0\,,
$$

or

$$
(10)\qquad \phi_{kt} + \phi_{kj}b^j = 0\,,
$$

or

$$
(11)\qquad \frac{d\phi_k}{dt} = 0\,.
$$

That is, the first derivatives of $\phi$ are zero on the deterministic trajectory (if they are zero initially). Turning to the quadratic terms, we have then

$$
(12)\qquad \phi_{tk\ell}\,\frac{\delta x^k\delta x^\ell}{2} + b^j_\ell\delta x^\ell\phi_{jk}\delta x^k + \frac{b^j}{2}\,\phi_{jk\ell}\delta x^k\delta x^\ell + \frac{a^{jm}}{2}\,\phi_{j\ell}\phi_{km}\delta x^\ell\delta x^k = 0\,.
$$

We can assume that the covariance matrix $(a^{jm})$ is symmetric. We have obtained a quadratic form in $\delta x$ which equals zero. From this we can conclude only that the symmetric part of the corresponding matrix vanishes. Therefore, we look at the symmetric part of the coefficients of $\delta x^k\delta x^\ell$ in (12). Only the second term requires symmetrization. Thus (12) implies that

$$
(13)\qquad \phi_{tk\ell} + b^j_\ell\phi_{jk} + b^j_k\phi_{j\ell} + b^j\phi_{jk\ell} + a^{jm}\phi_{j\ell}\phi_{km} = 0\,.
$$

As in the practice problem,

$$\phi_{k\ell t} + \phi_{k\ell j}\frac{dx^j}{dt} = \frac{d}{dt}\phi_{k\ell}\ . \tag{14}$$

Equation (13) can now be written in matrix form. Let

$$\begin{cases} \phi'' = (\phi_{\ell m})\ , \\ B = (b^j_\ell)\ , \\ A = (a^{jk})\ . \end{cases} \tag{15}$$

Then we have the matrix Ricatti equation

$$\frac{d}{dt}\phi'' + B\phi'' + \phi'' B^T + \phi'' A \phi'' = 0\ . \tag{16}$$

Let

$$W = (\phi'')^{-1}\ . \tag{17}$$

Then

$$\frac{dW}{dt} = -(\phi'')^{-1}\ (\frac{d}{dt}\phi'')\ (\phi'')^{-1} = -W\frac{d}{dt}\phi'' W\ . \tag{18}$$

After multiplication on the left and right by W, (16) becomes

$$\dot{W} = WB + B^T W + A\ . \tag{19}$$

In analogy with the previous case, $\frac{W}{N}$ is the covariance matrix.

Exercise 2. Solve (19) by the integrating factor method. Hint: Two integrating factors, left and right, are needed. If it were a scalar equation the factor would be $\exp(\int b)$. For the matrix equation this is not good enough. Try the solution C of $\dot{C} = B^T C$, and its transpose.

Thus, we have found an approximate solution, on the deterministic trajectory, of the form

$$V = e^{-1/2\delta x^T W^{-1}\delta x}\ z \tag{18}$$

where W is the covariance matrix. Noting from (1) that $\iint V_t dx$ is a divergence, we see that

$$\iint V dx = 1\ . \tag{19}$$

This determines z:

$$z = \frac{1}{[\det(2\pi W)]^{1/2}} \tag{20}$$

We can give A, B more specifically for equation (5) of 2.1.3. Let $s = x^1$, $r = x^2$.

Then

$$b^1 = -\,is = -\,s(1-s-r) = s(s+r-1)$$

$$b^2 = \rho i = \rho(1-s-r)$$

$$a^{11} = -\,b^1\ ,\quad a^{22} = b^2\ ,\quad a^{12} = a^{21} = 0$$

and

$$b_1^1 = 2s + r\ ,\quad b_1^2 = -\,\rho\ ,$$

$$b_2^1 = s\ ,\quad b_2^2 = -\,\rho\ .$$

Then

$$B = \begin{pmatrix} 2s + r & -\,\rho \\ s & -\,\rho \end{pmatrix}$$

$$A = \begin{pmatrix} s(1-s-r) & 0 \\ 0 & \rho(1-s-r) \end{pmatrix} .$$

No explicit solution of (19) seems to be available for this case.

Further details and applications are given in Ludwig (1974b).

## III. DIFFUSION EQUATIONS

### 1. Introduction to Diffusion Equations

The previous example suggests that it would be useful to take a closer look at diffusion equations. We do this now by way of several examples.

First example: Random walk.

Using the previous terminology, let

$$\text{Prob}[\text{Jump in } (t,t+\delta t)] = \beta\delta t + o(\delta t) \ .$$

Each jump has size $\delta x$. If a jump occurs, we assume

$$\text{Prob}[X \to X + \delta x] = 1/2 \ ,$$
$$\text{Prob}[X \to X - \delta x] = 1/2 \ .$$

Let $P_i(t) = \text{Prob}[X(t) = i\delta x]$. Then

$$P_i(t+\delta t) = (1-\beta\delta t)P_i(t) + \frac{1}{2}\beta\delta t(P_{i-1}(t) + P_{i+1}(t)) + o(\delta t) \ .$$

The limit of the difference quotient yields

$$\frac{dP_i(t)}{dt} = -\beta P_i(t) + \frac{\beta}{2}P_{i-1}(t) + \frac{\beta}{2}P_{i+1}(t) \ . \tag{1}$$

In preparation for the limit $\delta x \to o$, we let $v(t,x) = P_i(t)$. Then (1) may be rewritten as the difference equation

$$\frac{\partial v}{\partial t} = -\beta v(t,x) + \frac{\beta}{2}[v(t,x-\delta x) + v(t,x+\delta x)] \ .$$

The differences may be expanded by Taylor's Theorem, to obtain

$$\frac{\partial v}{\partial t} = -\frac{\beta}{2}v_x\delta x + \frac{\beta}{2}v_{xx}\frac{(\delta x)^2}{2} + \frac{\beta}{2}v_x\delta x + \frac{\beta}{2}v_{xx}\frac{(\delta x)^2}{2} + \ldots$$
$$= \frac{\beta}{2}v_{xx}(\delta x)^2 + \ldots \ .$$

Evidently, $\beta$ should be proportional to $1/(\delta x)^2$ if we are to obtain a limiting differential equation. Therefore, we shall set $a = \beta(\delta x)^2$, and we shall assume that $a$ is independent of $\delta x$. Then to leading order at least $v$ satisfies

(2) $$\frac{\partial v}{\partial t} = \frac{1}{2} a v_{xx} .$$

This calculation suggests that, in the limit, the density for X should satisfy (2).

On the other hand, X can be represented as a sum of independent, identically distributed random variables. The central limit theorem provides another description of the limiting behavior of X. Let Y be the random variable:

$$\text{Prob}[Y = \delta x] = 1/2 ,$$

$$\text{Prob}[Y = -\delta x] = 1/2 .$$

Then $E(Y) = 0$, and $E(Y^2) = (\delta x)^2$. The expected number of jumps made by X in the interval $(0,t)$ is $\beta t$. Thus $X(t)$ is approximately given by the sum of $\beta t$ independent random variables with the same distribution as Y so that

(3) $$E[X] = \beta t E[Y] = 0$$

(4) $$E[X^2] = \beta t E[Y^2] = \beta t(\delta x)^2 = at .$$

From the central limit theorem, we see that X should have an approximately normal distribution with mean 0, and variance at. Therefore the density v should be

(5) $$v = \frac{1}{\sqrt{2\pi a t}} e^{-\frac{x^2}{2at}} .$$

Exercise: Verify that v is a solution of $v_t = \frac{1}{2} a v_{xx}$, the heat equation.

This is a fundamental solution. This is seen by noting that it satisfies an initial condition that is a delta function; i.e., a point source.

Remark: This result suggests that sample paths of the diffusion process are limits of random walks and so, rather 'wiggly'. Note that (4) suggests that, in some sense

$$|X| \sim \sqrt{at}$$

and so, at $t = 0$,

$$\left|\frac{\delta X}{\delta t}\right| \sim \sqrt{\frac{a}{\delta t}} .$$

This expression doesn't have a finite limit as $\delta t \to 0$. Therefore X is probably not differentiable at t = 0 (or anywhere else, for that matter). But $|\delta X| \sim K\sqrt{\delta t}$ and so $\delta X$ is continuous. The sort of subtleties suggested in these approximate statements have been the subject of much study by probabilists. By comparison, little effort has been devoted to the more central problem of the validity of the diffusion equations.

Second example: Gambler's Ruin.

Reference: Feller, Vol. II, Chapter X.

To give this the classical motivation, suppose that a gambler starts betting with capital R, and let X(t) be tis total winnings at time t. We interpret the event $X(t) \leq -R$ as the gambler's ruin. To be more specific, if money is won or lost in discrete units and the probability of winning and losing are equal, then we can represent the process graphically by some polygonal path.

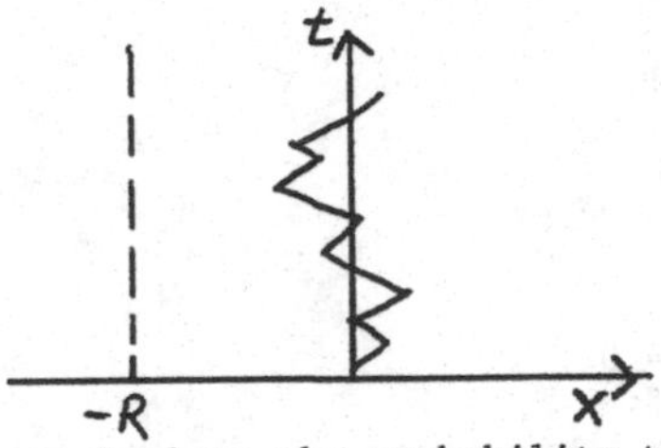


It is possible to calculate the probability that X(t) reaches -R (the probability of ruin) by using the symmetry of the basic transition probabilities. We note that, for any path which reaches -R, its reflection in -R is equally likely.

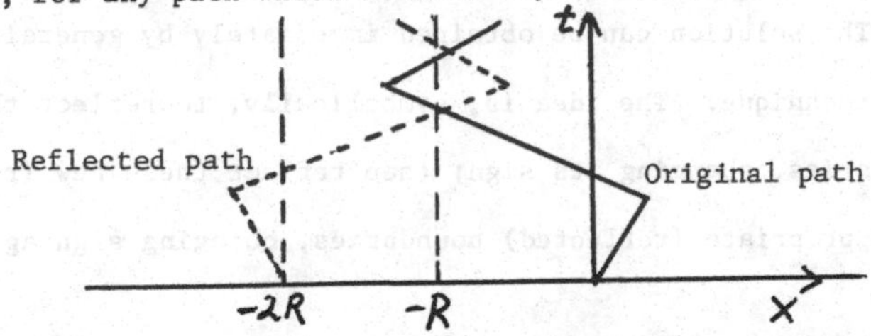


Let $v_R(t,x)$ be the conditional density of paths from the origin through (t,x) that have reached -R. Since for every path reaching -R from 0 there is an equally likely (reflected) path starting from -2R,

(6) $$v_R(t,x) = \frac{1}{\sqrt{2\pi at}} e^{-\frac{1}{2at}(x-(-2R))^2} .$$

Then

(7) $$\text{Prob}[x \le X(t) \le x + \delta x, \text{ and X did not reach R}] = (v(t,x) - v_R(t,x))dx$$
$$= \frac{dx}{\sqrt{2\pi at}} [e^{-\frac{1}{2at}x^2} - e^{-\frac{1}{2at}(x+2R)^2}] .$$

To obtain the probability Q(t) of not reaching -R by time t we integrate:

(8) $$Q(t) = \frac{1}{\sqrt{2\pi at}} \int_{-R}^{\infty} [e^{-\frac{1}{2at}x^2} - e^{-\frac{1}{2at}(x+2R)^2}]dx .$$

It is easily seen that Q goes to zero at $t \to \infty$.

Now, let us suppose that not only the gambler, but also the house has limited resources H. Then the process ends when X = -R or X = H. Now we have the problem

(9) $$v_t = \frac{1}{2} av_{xx}$$

with side conditions

(10) $$v(t,-R) = v(t,H) = 0 ,$$

(11) $$v(0,X) = \delta(x) ,$$

where $\delta(x)$ is the delta function. This is a clssical boundary value problem for the heat equation. The solution can be obtained immediately by generalizing the previous reflection technique. The idea is, symbolically, to reflect the source across the two boundaries, changing its sign; then reflect these new (reflected) sources across the appropriate (reflected) boundaries, changing sign again; and so on...

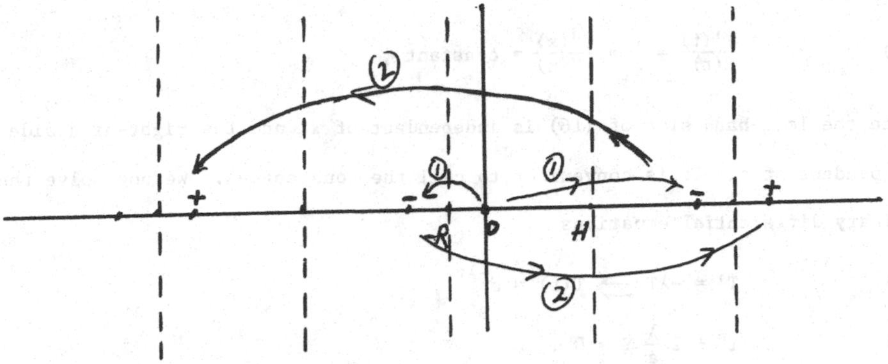


In this manner we formally obtain a sum of infinitely many terms that satisfies the boundary conditions. Successive terms of the sum (successive reflections) cancel the values of previous terms (reflections) at one of the boundaries. If t is small, the residual error drops off rapidly since the terms are of the form

$$\frac{1}{\sqrt{2\pi at}} e^{-\frac{1}{2at}(x-x_{image})^2} .$$

That is, for small t, higher terms in the expansion contribute little. As t increases these become more important and convergence becomes very slow. Therefore the image method works well for small t, but it is unsatisfactory for large t.

A second method, which works well for large t, is that of separation of variables, or Fourier series. For convenience we change the problem slightly to

(12) $$v_t = \frac{1}{2} av_{xx} ,$$

with side conditions:

(13) $$v(t,0) = v(t,1) = 0 ,$$

(14) $$v(0,x) = \delta(x-x_o) .$$

We assume a solution of the form

(15) $$v = T(t)Y(x) ,$$

i.e. we use the method of separation of variables. Then from the differential equation (12) we have

$$\frac{T'(t)}{T(t)} = \frac{1}{2}\, a\, \frac{Y''(x)}{Y(x)} = \text{constant}\ , \tag{16}$$

since the left-hand side of (16) is independent of x, and the right-hand side is independent of t. It is convenient to call the constant $-\lambda$. We now solve the two ordinary differential equations

$$T' = -\lambda T \Longrightarrow T(t) = e^{-\lambda t}\ , \tag{17}$$

$$Y'' + 2\,\frac{\lambda}{a}\, Y = 0\ . \tag{18}$$

The first boundary condition of (13) and (18) together imply that

$$Y = \sin \sqrt{\frac{2\lambda}{a}}\, x\ . \tag{19}$$

Then the second condition of (13) implies that

$$n\pi = \sqrt{\frac{2\lambda}{a}}\ , \quad \text{i.e. } \lambda = \frac{a}{2}\, n^2\pi^2\ . \tag{20}$$

Putting these results together, we see that there are solutions of (12) of the form

$$v_n = e^{-\frac{a}{2} n^2 \pi^2 t} \sin n\pi x\ . \tag{21}$$

A more general solution is of the form

$$v = \sum_{n=1}^{\infty} c_n v_n\ . \tag{22}$$

Now the condition (14) can be used to compute the coefficients $c_n$. We use the orthogonality relations

$$\int_0^1 \sin n\pi x \sin m\pi x dx = \begin{cases} \frac{1}{2} \text{ if } m = n \\ 0 \text{ otherwise}\ , \end{cases} \tag{23}$$

Now (22), (23) and (14) imply that

$$\int_0^1 \sin m\pi x \sum c_n \sin n\pi x\, dx = \int_0^1 \sin m\pi x\ \delta(x-x_0) dx\ ,$$

or

$$\int_0^1 c_m (\sin m\pi x)^2 dx = \sin m\pi x_o ,$$

or

(24) $$c_m = 2\sin m\pi x_o .$$

Putting things together, the solution of (12), (13) and (14) is

(25) $$v = \sum_n e^{-\frac{a}{2} n^2 \pi^2 t} \sin n\pi x_o \sin n\pi x .$$

Strictly speaking, such an expansion is not justified until the completeness of the set of eigenfunctions has been proven. For large t, only one term is required:

(26) $$v \sim e^{-\frac{a}{2}\pi^2 t} \sin \pi x_o \sin \pi x .$$

The method of images and the method of Fourier series are complementary, as was indicated above. It is in fact possible to go from one to the other by a transformation involving the Poisson sum formula (Jacobi's Theta transformation). Reference: Feller, p. 330.

## 2. Derivation of the Forward and Backward Diffusion Equations

Reference: Crow and Kimura (1970), Chapters 8 and 9.

Let X(t) be a process as before. Let $\delta X$ have the conditional density q, if $X(t) = x$. We shall think of q as an infinitesimal transition probability

(1) $$q(\delta t, x, \xi) d\xi = \text{Prob}[\xi \leq \delta X \leq \xi + d\xi | X(t) = x] .$$

Of course,

(2) $$\int q(\delta t, x, \xi) d\xi = 1 .$$

Then

(3) $$v(t+\delta t, x) = \int v(t, x-\xi) q(\delta t, x-\xi, \xi) d\xi + o(\delta t) ,$$

and (2) implies v may be represented in the form

$$(4) \qquad v(t,x) = \int v(t,x)q(\delta t,x,\xi)d\xi \ .$$

Therefore, after subtracting (4) from (3),

$$(5) \qquad \delta v = v(t+\delta t,x) - v(t,x) = \int [vq\big|_{t,x-\xi,\xi} - vq\big|_{t,x,\xi}]d\xi \ .$$

After expanding the integrand by Taylor series we obtain

$$(6) \qquad \delta v = \int [-\xi\partial_x(vq) + \frac{1}{2}\xi^2\partial_x^2(vq) + \ldots]d\xi \ ,$$

$$(7) \qquad \delta v = -\partial_x \int \xi(vq)\big|_{t,x,\xi} d\xi + \frac{1}{2}\partial_x^2 \int \xi^2 vq\big|_{t,x,\xi} d\xi + \ldots \ ,$$

$$(8) \qquad \delta v = -\partial_x[v(t,x)\int \xi q(\delta t,x,\xi)d\xi] + \frac{1}{2}\partial_x^2[v(t,x)\int \xi^2 q(\delta t,x,\xi) + \ldots \ .$$

Now, we make appropriate assumptions about the moments of q:

$$(9) \qquad E[\delta X|X(t) = x] = \int \xi q(\delta t,x,\xi)d\xi = b(x)\delta t + o(\delta t) \ ,$$

and

$$(10) \qquad E((\delta X)^2|X(t) = x) = \int \xi^2 q(\delta t,x,\xi)d\xi = a(x)\delta t + o(\delta t) \ .$$

Then we have that

$$(11) \qquad \delta v = \delta t[-\partial_x(bv) + \frac{1}{2}\partial_x^2(av)] \ , \text{ i.e.}$$

$$(12) \qquad \frac{\partial v}{\partial t} = -\frac{\partial}{\partial x}(bv) + \frac{1}{2}\left(\frac{\partial}{\partial x}\right)^2(av) \ .$$

Exercise: Derive a similar equation for N dimensions. (Let X be a vector).

We have derived the forward diffusion equation, also known as Kolmogorov's forward equation or the Fokker-Planck equation. We shall presently derive a closely related equation for the same process, the backward equation.

Before doing so, we shall illustrate (9) and (10) by means of a simple model from genetics. Consider two alleles $A_1$, $A_2$; i.e., two possible forms a given gene may take in a population of N diploid organisms (i.e. 2N chromosomes which contain one or the other of these alleles. Let Y(t) be the number of $A_1$ alleles, and let $X = \frac{Y}{2N}$. Now X, and also time, are discrete. But for large populations and a large number of generations, we may expect a continuous approximation to be adequate.

Assuming random mating (random sampling of alleles), and non-overlapping generations, we shall investigate the problem of random genetic drift. This problem was first studied extensively by the geneticist Sewall Wright.

According to our assumptions, Y(t+1) has a binomial distribution, and if Y(t) = 2Nx, then Y(t+1) has mean and variance

(13) $$E(Y(t+1)) = 2Nx(t) ,$$

(14) $$\sigma^2_{Y(t+1)} = 2Nx(1-x) .$$

These expectations and those which follow are conditional on Y(t) = 2Nx.

Let $\delta Y = Y(t+1) - Y(t)$, and $\delta t = 1$. Then (13) and (14) imply that

(15) $$E(\delta Y) = E(Y(t+1)) - 2Nx = 0 ,$$

and

(16) $$E((\delta Y)^2) = \sigma^2_{Y(t+1)} = 2Nx(1-x) .$$

Hence

(17) $$E((\delta X)^2 = \frac{1}{2N} x(1-x) .$$

Then, since $\delta t = 1$, (9) and (10) are satisfied with

(18) $$a(x) = \frac{1}{2N} x(1-x) , \quad b(x) = 0 .$$

## The Backward Equation

Now consider the density

(19) $$v(t,x,x_o)dx = \text{Prob}[x \leq X(t) \leq x + dx | X(0) = x_o] .$$

Neglecting higher moments, we have shown that v satisfies the forward equation (12) as a function of x. How does v depend on $x_o$? We have

$$v(t+\delta t,x,x_o) = \text{Prob}[X(\delta t) \text{ near } x_o+\xi] \cdot \text{Prob}[X(t+\delta t) \text{ near } x | X(\delta t) \text{ near } x_o+\xi] .$$

Thus

$$v(t+\delta t,x,x_o) = \int q(\delta t,x_o,\xi)v(t,x,x_o+\xi)d\xi \ . \tag{20}$$

By expanding v with respect to its last argument, we obtain

$$v(t,x,x_o+\xi) = v(t,x,x_o) + \xi \frac{\partial v}{\partial x_o}(t,x,x_o) + \frac{1}{2}\xi^2 \frac{\partial^2 v}{\partial x_o^2} + \dots , \tag{21}$$

$$v(t+\delta t,x,x_o) = \int qv d\xi + \int q\xi \frac{\partial v}{\partial x_o} d\xi + \frac{1}{2}\int q\xi^2 \frac{\partial^2 v}{\partial x_o^2} d\xi + \dots . \tag{22}$$

In view of (9) and (10), (22) becomes

$$v(t+\delta t,x,x_o) = v(t,x,x_o) + b(x_o)\frac{\partial v}{\partial x}\delta t + \frac{1}{2}a(x_o)\frac{\partial^2 v}{x_o^2}\delta t + \dots . \tag{23}$$

Then, dividing by $\delta t$ and passing to the limit as $t \to o$, we have the backward equation:

$$\frac{\partial v}{\partial t} = \frac{1}{2}a(x_o)\frac{\partial^2}{\partial x_o^2}v + b(x_o)\frac{\partial}{\partial x_o}v \ . \tag{24}$$

So $v(t,x,x_o)$ satisfies both forward and backward equations. Since it also satisfies the intial condition $v(0,x,x_o) = \delta(x-x_o)$, v is a fundamental solution of each equation.

## The Forward and Backward Operators as Adjoints

Now we examine an important property of equations (12) and (24). First we define the corresponding operators. Let

$$Lu = \frac{1}{2}a(x)\frac{\partial^2}{\partial x^2}u + b(x)\frac{\partial u}{\partial x} ,$$

$$L^*v = \frac{1}{2}\frac{\partial^2}{\partial x^2}(a(x)v) - \frac{\partial}{\partial x}(b(x)v) \ .$$

If boundary conditions of the proper sort are imposed, then L and $L^*$ are formal adjoints of each other, i.e.

$$\langle Lu,v \rangle = \langle u,L^*v \rangle , \text{ for all } u \text{ and } v , \tag{25}$$

where the inner product <u,v> is defined by $\int_{x_1}^{x_2} u(x)v(x)dx$. Frequently we shall take $x_1 = 0$, $x_2 = 1$.

In order to obtain (25), we integrate by parts as follows:

$$(26) \qquad \int_{x_1}^{x_2} v(\frac{1}{2} a \frac{\partial^2}{\partial x^2} u + b \frac{\partial u}{\partial x})dx = \int_{x_1}^{x_2} u(\frac{\partial^2}{\partial x^2} (av) - \frac{\partial}{\partial x} (bv))dx + \text{boundary terms}$$

If the boundary terms vanish, then (26) has exactly the form (25).

Exercise: Show that the boundary terms are

$$(27) \qquad [\frac{\partial u}{\partial x} (\frac{1}{2} av) - u(\frac{\partial}{\partial x}\left(\frac{1}{2} av\right) - bv)] .$$

We shall impose boundary conditions on u and/or v so that these boundary terms vanish. Remark: Note that L (or $L^*$) is not self-adjoint: $L \neq L^*$. This is different from the situation frequently encountered in mathematical physics. The non-self-adjointness of these operators is intimately connected with the fact that the points x and $x_o$ cannot be interchanged, i.e. "reciprocity" is violated.

## Eigenfunctions and Orthogonality

Suppose we have boundary conditions such that $\langle Lu,v\rangle = \langle u,L^*v\rangle$. Let u, v be functions satisfying the boundary conditions such that $Lu = -\lambda u$, $L^*v = -\mu v$, for constants $\lambda$ and $\mu$. Then u and v are called eigenfunctions of L and $L^*$ respectively. The numbers $\lambda$ and $\mu$ are the corresponding eigenvalues. From linear algebra, we recall the following Theorem: $\lambda \neq \mu \Longrightarrow u \perp v$.

Proof:

$$-\lambda \langle u,v\rangle = \langle Lu,v\rangle = \langle u,L^*v\rangle = \langle u,-\mu v\rangle = -\mu \langle u,v\rangle.$$

If $\lambda = \mu$, we have no information on <u,v>, but if $\lambda \neq \mu$, then clearly $\langle u,v\rangle = 0$, or $u \perp v$.

## 3. Random Genetic Drift

We have seen that in this case, from (2.18), b = o, a = x(1-x)/2N. We take the basic interval as [0,1], since $X = \frac{Y}{2N}$, and $0 \leq Y \leq 2N$. Note that a = 0 at the end points. Since a is the coefficient of the highest derivative in (2.12) or (2.24), these equations are singular at the end points. The standard theory of parabolic partial different equations does not apply to such cases, and therefore we must use special methods. Since $b = 0$, the boundary terms reduce to

(1) $$\left[-\frac{1}{2}uv\frac{\partial a}{\partial x}\right]_o^1 .$$

Now we must determine the boundary conditions. $v(t,x,x_o)$ is the probability density for $X(t)$ at $x$ if $X(0) = x_o$. We know that if $x_o = 0$, then $v(t,x,0) = 0$ for $x > 0$, and if $x_o = 1$, then $v(t,x,1) = 0$ for $x < 1$. So the appropriate boundary conditions for the forward equation are

(2) $$u(0) = u(1) = 0 .$$

If these conditions are satisfied, then (1) will vanish, without imposing any condition on v, other than regularity at the end points. Now suppose we know eigenfunctions for L and $L^*$. (These are actually calculated in Crow and Kimura (1970)). By separating variables, we are led to try a solution of the form

(3) $$v(t,x,x_o) = \sum_n c_n v_n(x) e^{-\lambda_n t} ,$$

where the $v_n$ are eigenfunctions of $L^*$,

(4) $$L^* v_n = -\lambda_n v_n$$

Then for $t = 0$,

(5) $$\delta(x-x_o) = v(0,x,x_o) = \sum_n c_n v_n(x) .$$

We multiply (5) by $u_m(x)$, which satisfies

(6) $$Lu_m = -\lambda_m u_m ,$$

and boundary conditions. Because of the orthogonality relation, this yields

(7) $$u_m(x_o) = \int_o^1 \delta(x-x_o)u_m(x)dx = \sum_n c_n \int_o^1 v_n(x)u_m(x)dx = c_m \int_o^1 v_m(x)u_m(x)dx ,$$

or

$$(8)\qquad c_m = \frac{u_m(x_o)}{\int_0^1 u_m(x)v_m(x)dx} .$$

If we order the eigenvalues so that $\lambda_1$ is the smallest, then for large t only one term of the expansion is significant, and

$$(9)\qquad v(t,x,x_o) \sim \frac{u_1(x_o)v_1(x)e^{-\lambda_1 t}}{\int u_1 v_1 dx} .$$

From the definition of $L^*$, $v_1$ is supposed to satisfy

$$(10)\qquad L^* v = \frac{\partial^2}{\partial x^2}\left(\frac{1}{4N} x(1-x)v\right) = -\lambda_1 v .$$

Note that if v is a polynomial of degree n, then $L^* v$ is a polynomial of degree n. The eigenfunction which corresponds to the smallest eigenvalue should not vanish in the interval (0,1).

So we try v = 1. This gives

$$\frac{\partial^2}{\partial x^2} \frac{1}{4N} x(1-x) = -\frac{1}{2N} \cdot 1 ,$$

or

$$(11)\qquad v_1 = 1 , \quad \lambda_1 = \frac{1}{2N} .$$

$u_1$ is supposed to satisfy

$$(12)\qquad \frac{1}{4N} x(1-x) \frac{\partial^2}{\partial x^2} u_1 = -\lambda_1 u_1 ,$$

$$(13)\qquad u_1(0) = 0 , \quad u_1(1) = 1 .$$

Clearly if $u_1$ is a polynomial, it is of degree $\geq 2$. So we try $u_1 = x(1-x)$. This gives

$$(14)\qquad u_1 = x(1-x) , \quad \lambda_1 = \frac{1}{2N}$$

Finally, we now compute the inner product

$$\int_0^1 u_1 v_1 dx = \int_0^1 x(1-x)dx = \frac{1}{6} .$$

Thus

(15) $$v(t,x,x_o) \sim 6x_o(1-x_o)e^{-t/2N} .$$

From this expression we see that v is independent of x for large t.

Further terms in the series (3) are given in Crow and Kimura (1970). From (15) we conjecture that, if one allele is not fixed (if x does not reach 0 or 1) fairly quickly, then the gene ratio is uniformly distributed. Also, we would expect that, for genes not fixed rapidly, the expected time to fixation will be of order 2N generations.

We can now answer two natural questions:

1. What is the probability $u(t,x_o)$ of no fixation by time t, if $X(0) = x_o$?

Answer:

(16) $$u(t,x_o) = \int_0^1 v(t,x,x_o)dx .$$

Since v satisfies the backward equation (2.24) as a function of $x_o$, and linear combinations of solutions of linear equation are also solutions, we consider the integral as a limit of a sum and conclude that u given by (16) is also a solution of the backward equation:

(17) $$\frac{\partial u}{\partial t} = x_o(1-x_o) \frac{\partial^2}{\partial x_o^2} u .$$

It is clear that u must also satisfy the initial condition:

(18) $$u(o,x_o) = \int_0^1 \delta(x-x_o)dx = 1 \text{ if } 0 < x_o < 1.$$

The solution u must also satisfy the boundary conditions

(19) $$u(t,o) = u(t,1) = o ,$$

because at x = 0 or 1 we would have fixation from the beginning. The complete solution of (17), (18) and (19) would be rather complicated. However, from (15) and (16) we now know that, for large t,

$$u \sim 6x_o(1-x_o)e^{-\frac{1}{2N}t} \quad . \tag{20}$$

The behavior of u for small t will be considered in the next section.

The present methods enable us to answer a refinement of the first question:

2. What is the probability of fixation (absorption) at x = 1, say, by time t?

Now let $u(t,x_o)$ = probability of fixation of first allele, by time t, starting at $x_o$. Then, as before, u is a limit of linear combinations of solutions of (2.24) (in this case, a complementary set of solutions that reach fixation by time t), and hence u satisfies (17). The appropriate side conditions in this case are

I. C. (21) $u(0,x_o) = o$ if $0 \leq x_o < 1$

B. C. (22) $u(t,o) = 0 \ , \quad u(t,1) = 1 \ .$

With these conditions we then solve (17). Now, for large enough t we can set $u_t = 0$ and ignore the initial conditions. We then have the problem

$$\frac{\partial^2 u}{\partial x_o^2} = 0 \ ,$$

$$u(t,0) = 0$$

$$u(t,1) = 1 \ .$$

The solution to this is $u = x_o$, so the probability of eventual fixation of an allele is given by its initial proportion in the population.

## 4. Solutions which are valid for small time

In order to complete the picture we now investigate the nature of the solution for small t. The method of images which was used in Section 1 suggests that the solution takes the form of a Gaussian about some mean (deterministic) value. By our earlier work (II. 4) , this mean should satisfy

(1) $$\frac{dx}{dt} = b(x) = 0 ,$$

so the deterministic solution is

(2) $$x_d(t) = x_o .$$

According to (II.4.19) the variance w satisfies

(3) $$\frac{dw(t)}{dt} = a(x) + \dots = \frac{x(1-x)}{2N}$$

So, for small t,

(4) $$w \sim \frac{x_o(1-x_o)t}{2N} ,$$

and the approximate solution becomes

(5) $$v \sim \frac{1}{\sqrt{\frac{\pi x_o(1-x_o)}{N}}} \exp[-\frac{N(x-x_o)^2}{x_o(1-x_o)t}] .$$

A slightly better procedure is the following: we shall assume a solution of the form

(6) $$v = e^{-2N\phi(x,t)}z(x,t) .$$

Then

$$v_t = -2N\phi_t v + \dots ,$$

$$v_x = -2N\phi_x v + \dots ,$$

$$v_{xx} = 4N^2\phi_x^2 v + \dots .$$

Putting these into the forward equation

(7) $$v_t = \frac{1}{4N}\frac{\partial^2}{\partial x^2}(x(1-x)v) ,$$

we have

(8) $$-2N\phi_t v = \frac{1}{4N} x(1-x) \cdot 4N^2\phi_x{}^2 v + \dots .$$

Therefore $\phi$ must satisfy

$$(9) \qquad -\phi_t = \frac{x(1-x)}{2}\phi_x^2 = \frac{1}{2}(\sqrt{x(1-x)}\ \phi_x)^2 .$$

The variable coefficient on the right-hand side of (9) can be removed by a change of variables. If

$$(10) \qquad d\xi = \frac{dx}{\sqrt{x(1-x)}} ,$$

then

$$(11) \qquad \frac{\partial\phi}{\partial\xi} = \sqrt{x(1-x)}\ \frac{\partial\phi}{\partial x} .$$

We now have the equation

$$(12) \qquad \phi_t + \frac{1}{2}\phi_\xi^2 = 0 .$$

By analogy with the solution of the heat equation (1.5),

$$(13) \qquad \phi = \frac{1}{2}\frac{(\xi-\xi_0)^2}{t}$$

And so the solution (6) has the form,

$$(14) \qquad v \sim \exp[-\frac{N(\xi-\xi_0)^2}{t}]z(t,x) ,$$

where z must be determined by normalization, so that

$$\int vdx = 1.$$

The qualitative behavior of the solution of the random drift problem can be understood fairly completely on the basis of (14) and the formulas of the previous section. The exact solutions are graphed in Crow and Kimura (1970). For $x_0 = 1/2$, the graphs are of the form

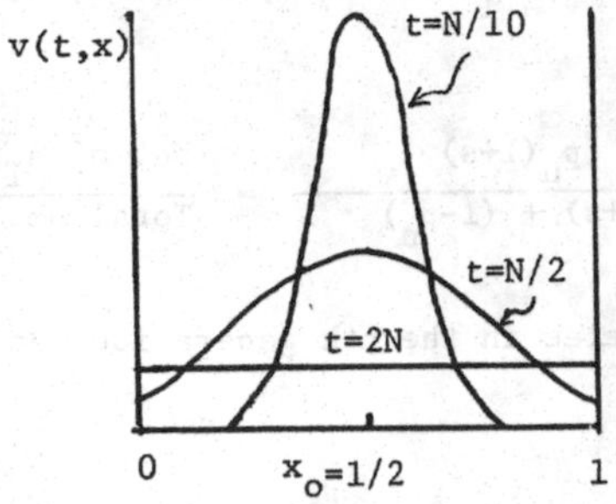

We see that by the time t = N/2 there is considerable leakage or absorption at the boundaries. As expected, by the time T = 2N, the distribution has approximately the large time form

$$v = 6x_o(1-x_o)e^{-t/2N}$$

and is essentially independent of x.

For an asymmetric initial condition, $x_0 = 1/10$, we have:

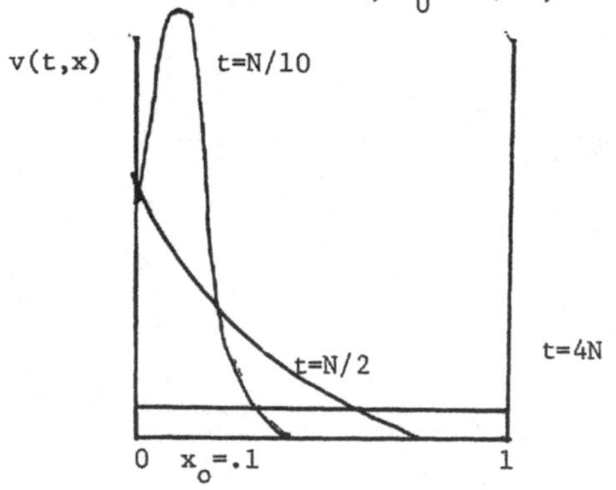


Here, the symmetry of the short-time approximation appears at t = N/10. At t = N/2 the greater absorption at the left boundary causes asymmetry. By t = 4N though, the symmetric large-time solution prevails. Further details and examples are given in Voronka and Keller (1975).

## 5. Random Drift and Selection

We are now ready to add a touch of realism, by letting one allele be a little better than another, in the evolutionary sense. For simplicity, we shall consider the haploid case. Suppose allele $A_1$ has a slight advantage over $A_2$. Let $p_n$ be the proportion of $A_1$ alleles in the gene-pool at generation n. Then, representing the relative selective advantage of $A_1$ over $A_2$ by (1+s):1, we have for the next generation

(1) $$p_{n+1} = \frac{p_n(1+s)}{p_n(1+s) + (1-p_n) \cdot 1} = \frac{\text{No. of } A_1 \text{ alleles}}{\text{Total no. of alleles}}$$

Let Y(n) = number of $A_1$ alleles in the nth generation, and let X(n) = Y(n)/N.

Then

$$E[Y(n+1)|X(n) = x] = N \frac{x(1+s)}{x(1+s) + 1-x} = N \frac{x(1+s)}{1+sx} .$$

Let $\delta Y = Y(n+1) - Y(n)$. Then

$$E[\delta Y|X(n) = x] = N \frac{sx(1-x)}{1+sx} .$$

Now let s be small (a good approximation for some, but not all cases that have been studied by geneticists.) Then

(2) $$E[\delta Y|X(n) = x] = Nsx(1-x) ,$$

(3) $$E[(\delta Y)^2|X(n) = x] = Nx(1-x) ,$$

(4) $$E[\delta X|X(n) = x] = sx(1-x),$$

(5) $$E[(\delta X)^2|X(n) = x] = \frac{1}{N} x(1-x) .$$

From (2.9 - 2.12), we have the forward equation

(6) $$v_t = - \frac{\partial}{\partial x} (sx(1-x)v) + \frac{1}{2N} \frac{\partial^2}{\partial x^2} (x(1-x)v) .$$

We could obtain large time solutions for this equation by performing an eigenfunction expansion (see Crow and Kimura (1970)). Instead we shall start with the simpler problem of finding the probability of fixation. The corresponding backward equation is

(7) $$u_t = sx_o(1-x_o) \frac{\partial u}{\partial x_o} + \frac{1}{2N} x_o(1-x_o) \frac{\partial^2 u}{\partial x_o^2} .$$

The appropriate side conditions are

(8) $$u(o,x_o) = o , \quad o \leq x_o < 1 ,$$

(9) $$u(t,o) = o , \quad u(t,1) = 1 ,$$

if, as before, we let u be the probability of fixation of $A_1$ by generation t. As before, for large t we set $u_t = 0$ and ignore the initial conditions to obtain the easier problem

(10) $$sx_o(1-x_o)\frac{\partial u}{\partial x_o} + \frac{1}{2N}x_o(1-x_o)\frac{\partial^2 u}{\partial x_o^2} = 0 .$$

Letting $u' = \frac{u}{x_o}$, this gives

(11) $$\frac{u''}{u'} = -2Ns ,$$

(12) $$u' = \text{const.} \cdot \exp[-2Nsx_o] .$$

After integrating (12) and applying (9),

(13) $$u = \frac{1-e^{-2Nsx_o}}{1-e^{-2Ns}}$$

Now let Ns be large. In this case (which may obtain in some natural populations) the graph of the solution is of the form

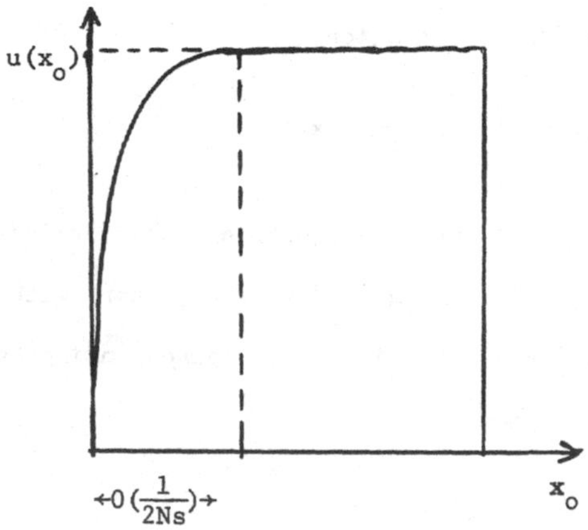


The sharp change at the left end point is called a boundary layer; its width decreases as Ns increases.

Exercise: The foregoing development can be adapted to diploid organisms (without dominance) by replacing N by 2N throughout. In the case of dominance, it will be shown later that the appropriate coefficient for the selection term is

(14) $$b(x) = x(1-x)s[h+(1-2h)x] .$$

Here h represents the effect of the dominance. The previous case corresponds to $h = \frac{1}{2}$ (no domiance), and h = 1 corresponds to complete dominance. Derive the analogous large time solution, using this expression for b.

We now consider the solution at the boundary layer. Let $Y(0) = \ell$, i.e. $x_o = \frac{\ell}{N}$. Then if Ns > > 1, the large time solution (13) is approximately

$$u \sim 1 - e^{-2s\ell} . \tag{15}$$

## Solution by Branching Process

The validity of the diffusion equation for small $x_o$ is not obvious, since it rests on replacing an integer-valued variable by a continuous variable. In order to check this point, we shall compute the probability of extinction of the $A_1$ allele by means of a branching process.

We shall assume that the number of descendents of a single individual has a Poisson distribution, with mean m. Then the extinction probability ρ for the family founded by a single individual is given by Exercise 2 on p. 22 as

$$\rho = e^{m(\rho-1)}, \tag{16}$$

The probability of extinction q for the set of ℓ families founded by ℓ individuals is given by

$$q = \rho^{\ell} . \tag{17}$$

The result (15), on the other hand, suggests that

$$q = e^{-2s\ell} . \tag{18}$$

These two results agree if we can show that log ρ = -2s.

Since the allele $A_1$ has selective advantage s, we set

$$m = 1 + s . \tag{19}$$

If $-\mu = \log \rho$, (16) and (19) imply that

(20) $$-\mu = (1+s)(e^{-\mu}-1) .$$

If s is small, $\mu$ must also be small, since $\rho$ is close to 1. Therefore, to leading order, (20) becomes

(21) $$\begin{aligned} -\mu &= (1+s)(-\mu + \frac{1}{2}\mu^2 + \ldots) , \\ &= -\mu - s\mu + \frac{1}{2}\mu^2 + \ldots , \end{aligned}$$

and hence

(22) $$\mu \sim 2s .$$

We conclude that, if s is small, the results (15) and (17) agree. We have already imposed the condition that s be small in the derivation of the diffusion equation. Thus we are led to the conjecture that the diffusion equation is valid in the limit as $s \to o$. Such theorem have been proven by Feller (1951) and Watterson (1962).

## 6. Equilibrium Distributions—Wright's Formula

Reference: Crow and Kimura (1970), Chapter 9.

We now look for equilibrium solutions of the forward equation. We start by rewriting it in the form

(1) $$\partial_t v + \partial_x F = 0 ,$$

where

(2) $$F(t,x) = -\frac{1}{4N}\partial_x(av) + bv .$$

In order to interpret F, we integrate (1) over an arbitrary interval $[x_1,x_2]$. Thus

(3) $$\partial_t \int_{x_1}^{x_2} v dx = -[F(t,x)]_{x_1}^{x_2} .$$

That is, the rate of change of $\mathrm{Prob}[x_1 \leq X(t) \leq x_2]$ is given by the difference in the values of F at the end points of the interval. This suggests that F(t,x) represents the net probability flux at the point x at time t. The term bv in (2) clearly has this character; it represents a (deterministic) motion with velocity b. The first term on the left-hand side of (2) represents the net diffusion across x, since it is proportional to the difference of $\frac{1}{4N}$ av on either side of x.

If we set $\partial_t v = 0$ (at equilibrium), then (1) implies that

(4) $$F(t,x) = \text{const.}$$

Unless, by some mechanism, the process is continually being started at one end point and terminated at the other, the flux must vanish. Therefore we set F = 0, i.e.,

(5) $$\frac{1}{4N}\,\partial_x(av) = bv\ .$$

This equation can be interpreted as a balance between deterministic forces and diffusion effects. It is therefore a reasonable equation for equilibrium. From (5), we obtain

(6) $$\frac{1}{av}\,\partial_x(av) = 4N\,\frac{b}{a}\ ,$$

or

(7) $$v = \frac{c}{a(x)}\exp[4N\int^x \frac{b}{a}\,dx']\ ,$$

which is Wright's formula. The constant c is determined by the normalization condition

(8) $$\int v dx = 1.$$

In the case of random genetic drift, where b(x) = 0, and a(x) = x(1-x), we have

(9) $$v = \frac{c}{x(1-x)}\ .$$

This is not integrable over [0,1] for c ≠ 0, reflecting the fact that there is no equilibrium solution except total fixation. On the other hand, restricting ourselves to some closed sub-interval $[\varepsilon, 1-\varepsilon]$, we can regard this as a formula for the conditional distribution where neither allele has been fixed. This suggestion (see p. 129 Moran (1962)), allows us to integrate over the sub-interval to obtain c. Such a conditional distribution might be realized if there is a steady source of the alleles in the population. Thus, we are led naturally to incorporate mutation in the model. This modification will allow us to deal with the whole interval unambiguously.

Let

$$\mu = \text{rate of mutation of } A_1 \longrightarrow A_2$$

$$\nu = \text{rate of mutation of } A_2 \longrightarrow A_1$$

$$x = \text{proportion of } A_1 \text{ alleles .}$$

Then

$$\frac{dx}{dt} = -\mu x + \nu(1-x) + \text{selection terms .} \tag{10}$$

Assuming no selection,

$$b(x) = -\mu x + \nu(1-x) \ , \tag{11}$$

$$\frac{b}{a} = -\frac{\mu}{1-x} + \frac{\nu}{x} \ , \tag{12}$$

$$\int \frac{b}{a}\, dx = \mu \log(1-x) + \nu \log x \ , \tag{13}$$

and

$$v = cx^{4N\nu-1}(1-x)^{4N\mu-1} \ , \tag{14}$$

which is integrable, provided $\nu, \mu > 0$. If $4N\nu$, $4N\mu$ are small, (as they usually are), most of the mass is near the endpoints 0,1 and the minimum of v is at the point where

$$\frac{4N\nu-1}{x} = \frac{4N\mu-1}{1-x} \ .$$

We now consider the case when selection is not zero. The basic assumptions can be displayed conveniently in a chart:

| Genotype | $A_1A_1$ | $A_1A_2$ | $A_2A_2$ |
|---|---|---|---|
| Proportion in population (Hardy-Weinberg) | $p^2$ | $2pq$ | $q^2$ |
| Fitness | $1 + s$ | $1 + sh$ | $1$ |
| Number of $A_1$ gametes contributed | $2p^2(1 + s)$ | $2pq(1 + sh)$ | $0$ |
| Total number of gametes | $2p^2(1 + s)$ | $4pq(1 + sh)$ | $2q^2$ |

From this chart, we can obtain the rate of change of proportions per generation. If $p = p_n$ refers to the nth generation,

$$(15)\qquad p_{n+1} = \frac{2p^2(1+s) + 2pq(1+sh)}{2p^2(1+s) + 4pq(1+sh) + 2q^2}$$

$$= \frac{p + sp(p+hq)}{1 + sp(p+2hq)} ,$$

$$(16)\qquad p_{n+1} - p = \frac{p + sp(p+hq) - p - sp^2(p+2hq)}{1 + sp(p+2hq)} .$$

We assume now that change in a single generation, and hence s, is small. Otherwise the continuous model would be inappropriate. Then

$$(17)\qquad p_{n+1} - p \sim sp[p + hq - p^2 - 2pqh]$$

$$= spq[p + h - 2ph]$$

$$= sp(1-p)[h + p(1-2h)] .$$

We note that:

1. When $h = 1$, $A_1$ is dominant.

2. When $h = 0$, $A_2$ is dominant.

3. When $h = \frac{1}{2}$, there is no dominance.

4. When $h > 1$, there is heterosis, or heterozygote advantage, as in the well known case of blood type S(sickle-cell).

5. When $h < 0$, there is heterozygote disadvantage, as in the centrifugal selection of early speciation.

## The Case of Heterozygote Advantage

Now, the deterministic equation corresponding to (17) is (replacing p by x)

$$b(x) = \frac{dx}{dt} = sx(1-x)[h + x(1-2h)] \quad . \tag{18}$$

This is zero when $h + x(1 - 2h) = 0$, and so the non-trivial deterministic equilibrium occurs at

$$\hat{x} = \frac{h}{2h-1} \quad , \quad \text{if } h > 1 \; . \tag{19}$$

Since near 0, $b > 0$ and near 1, $b < 0$, this is a stable equilibrium. We see then that deterministically $x \to \hat{x}$ as $t \to \infty$. This is in striking constrast to the previous theories, which send x to the endpoints.

Incorporating the selection terms in our equation, we now have

$$\frac{b}{a} = -\frac{\mu}{1-x} + \frac{\nu}{x} + s(h + x(1-2h)) \; ,$$

$$= \frac{-\mu}{1-x} + \frac{\nu}{x} + s(1-2h)(x-\hat{x}) \; ,$$

$$\int \frac{b}{a}\, dx = \mu \log(1-x) + \nu \log x + s(1-2h)\, \frac{(x-\hat{x})^2}{2} \quad .$$

From (7),

(20) $$cx^{4N\nu-1}(1-x)^{4N\mu-1}\exp[4Ns(1-2h)\ \frac{(x-\hat{x})^2}{2}]\ .$$

If 4Ns(2h-1) is large, this looks like a Gaussian centered at $\hat{x}$ with variance $\frac{1}{8N(h-\frac{1}{2})s}$. At the endpoints it blows up, but in extremely narrow regions. Thus most of the probability mass lies about $\hat{x}$, but there is some small probability of quasi-fixation at the endpoints.

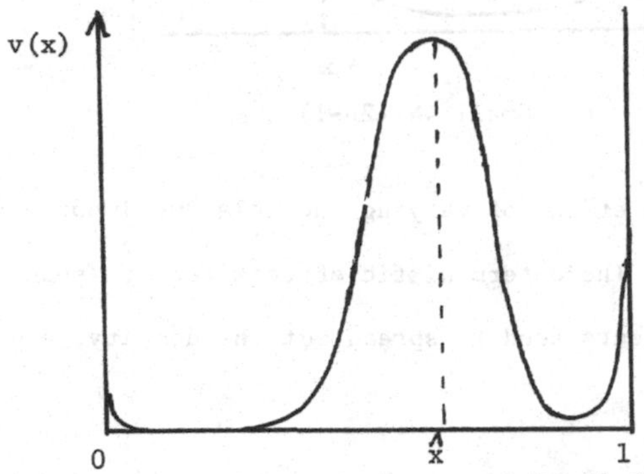


Equilibrium for large values of 4Ns(2h-1)

For lower values of 4Ns(2h-1), the equilibrium curve tends towards that of the zero selection case.

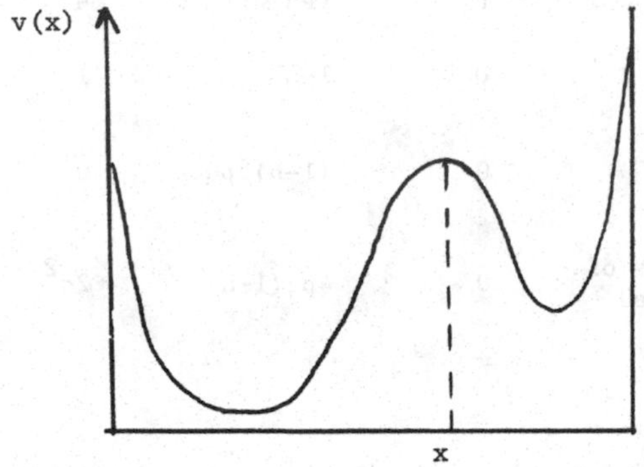


Smaller 4Ns(2h-1)

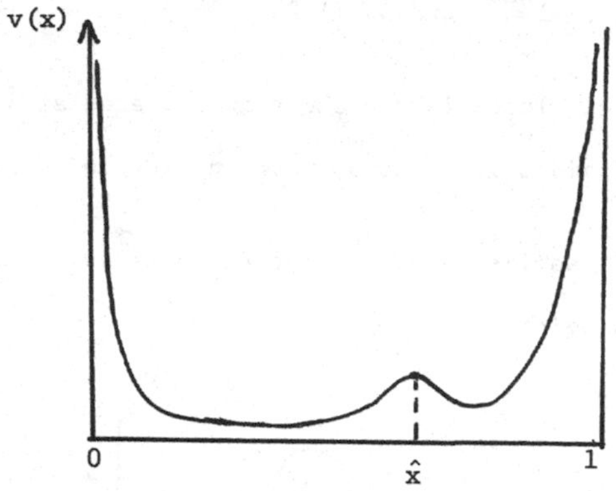


Small 4Ns(2h-1)

These curves illustrate the effect of varying the relative importance of stochastic and deterministic effects. The deterministic effects tend to peak the density near $\hat{x}$, while the stochastic effects tend to spread out the density, and sweep it into the endpoints.

## Lethal Recessives

As a furtherrexample, we apply Wright's theory to a case of extreme selection which is medically important. For a lethal recessive gene $A_1$ we have the table

| | $A_1A_1$ | $A_1A_2$ | $A_2A_2$ |
|---|---|---|---|
| frequency | $p^2$ | 2pq | $q^2$ |
| fitness | 0 | 1-h | 1 |
| number of $A_1$ gametes contributed | 0 | (1-h)2pq | 0 |
| number of both kinds of gametes contributed | 0 | 4pq(1-h) | $2q^2$ |

Then with $p = p_n$, $q = 1-p$

$$p_{n+1} - p = \frac{pq(1-h) - p(2pq(1-h) + q^2)}{q^2 + 2pq(1-h)} = \frac{p(1-h) - 2p^2(1-h) - pq}{q + 2p(1-h)} . \tag{21}$$

If p is small, which is usual for a lethal recessive, the denominator is near 1, and

$$(22) \qquad p_{n+1} - p \sim p - ph - 2p^2 + 2p^2h - p + p^2$$

$$= -p^2 - ph + 2p^2h$$

$$\sim -p^2 - ph \text{ , for small } h \text{ .}$$

The accuracy of this model can be improved by introducing the inbreeding coefficient, f. This is the probability that two homologous loci have the same allele, by descent from a common ancestor. Then in the table above we will have additional terms due to inbreeding, which are important for small p:

| | $A_1A_1$ | $A_1A_2$ | $A_2A_2$ |
|---|---|---|---|
| frequency | $p^2(1-f) + pf$ | $2pq(1-f)$ | $q^2(1-f) + qf$ |

If the resulting terms involving f are taken into account, and mutations are introduced, then

$$(23) \qquad \delta p = b = -p^2 - p(h+f) - \mu p + \nu(1-p) \text{ .}$$

As before

$$(24) \qquad a = p(1 - p) \text{ .}$$

Neglecting terms of higher order in p, this yields

$$(25) \qquad b/a = -p - (h+f) + \frac{\nu}{p} \text{ .}$$

The first two terms on the right are losses. The third is a source term representing new alleles coming from mutation. Then

$$(26) \qquad \int \frac{b}{a}\, dp = - \frac{1}{2} p^2 - (h+f)p + \nu \log p \text{ ,}$$

and Wright's formula yields

$$(27) \qquad v = cp^{4N\nu-1}\exp[-4N(\frac{1}{2} p^2 + (h+f)p)] \text{ .}$$

If we assume $p << (h+f)$, this is approximately a Gaussian density. The mean yields the result of deterministic theory:

(28) $$\bar{p} = \frac{\nu}{h+f} \quad ,$$

with variance

(29) $$\sigma_p^2 = \frac{\nu}{4N(h+f)^2} \; .$$

If $\sigma << \bar{p}$, this looks Gaussian near $\bar{p}$

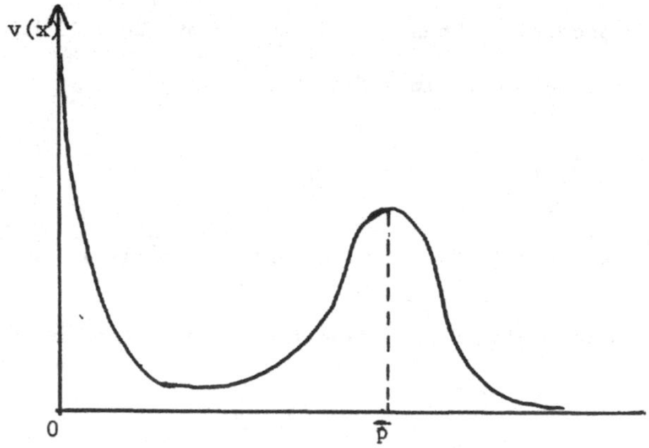


As $\sigma^2 \to 0$ the mass concentrates at $\bar{p}$, where gains due to mutations, $\nu$, balance the losses $(h+f)p$. For large N, this will be so, but for small N the singularity at 0 will dominate.

## IV. DYNAMICAL SYSTEMS PERTURBED BY NOISE

### 1. One Species

The growth of many populations (e.g. insects) is strongly influenced by the weather. Therefore, one cannot predict the growth of such a population without prior knowledge of what the weather will be. Symbolically,

(1) $$\frac{dX}{dt} = F(X,R) ,$$

where R is a function of t which represents the effects of the weather. If a number of populations are observed at different times or different places, then R will be a different function in each case. Since R cannot be predicted in any case, the problem appears to be unsolvable. However, certain statistics of the weather can be predicted from past data. This suggests that R be considered to be a random process whose moments, say, are known. One would like to describe the distribution of X in terms of the moments of R.

In general, this problem is still too difficult. Therefore we shall apply a perturbation procedure, i.e. consider R with a small random part:

(2) $$R = \bar{R}(X) + r$$

where $\bar{R}(X)$ is not random, and r is random with mean 0. We shall also assume that r is small and its distribution depends only on X(t). That is,

(3) $$\bar{R}(X) = E(R|X = x) ,$$

(4) $$E(r|X = x) = 0 ,$$

(5) $$E(r^2|X = x) = \varepsilon\,\alpha(x) ,$$

where ε is small, and α(x) remains to be specified. Note that we are assuming that X is Markov—that is, for any s, its distribution for t > s is independent of the previous history of X and R (the values of X and R for t < s). Then, by expanding F in a Taylor series about $\bar{R}$, (1) becomes

(6) $$\frac{dX}{dt} = F(X,\bar{R}) + \frac{\partial F}{\partial R}(X,\bar{R})r + \ldots .$$

Let

$$b(x) = F(x,\bar{R}(x)) .$$

Then

(7) $$dX = b(X)dt + \frac{\partial F}{\partial R} rdt ,$$

(8) $$E(dX|X = x) = b(x)dt ,$$

and the variance of dX is

(9) $$\sigma^2_{dX} = \left(\frac{\partial F}{\partial R}\right)^2 \varepsilon\alpha(x)dt^2$$

The analogy with (III.2.9) and (III.2.10) suggests that the right-hand side of (9) should be proportional to dt. If we define a new variable dS by

(1 ) $$dS = rdt ,$$

Then

(11) $$E(dS|X = x) = 0 ,$$

and instead of (5), we require that

(12) $$\sigma^2_{dS} = \varepsilon\gamma(x)dt .$$

For instance dS might be a Gaussian process as described in III.1. Then (9) becomes

(13) $$\sigma^2_{dX} = \left(\frac{\partial F}{\partial R}\right)^2 \varepsilon\gamma(x)dt = \varepsilon a(x)dt .$$

Let v be the density for X. Then the derivation of III.2 implies that

(14) $$v_t = \frac{\varepsilon}{2}\,\partial_x^2(a(x)v) - \partial_x(b(x)v)\ .$$

Example: Consider the logistic equation

(15) $$\frac{dx}{dt} = r'x(1 - \frac{x}{K})\ ,$$

where r' is the growth rate, and K is the carrying capacity. Now if we let r' or K be random the equation becomes more complicated. For instance, let

(16) $$r' = \bar{r} + \frac{dS}{dt}$$

Then

(17) $$dX = \bar{r}X(1 - \frac{X}{K})dt + X(1 - \frac{X}{K})dS\ ,$$

(18) $$E(dX|X = x) = \bar{r}x(1 - \frac{x}{K})$$

(19) $$\sigma_{dX}^2 = x^2(1 - \frac{x}{K})^2\varepsilon \quad \text{(letting } \gamma = 1)\ .$$

Thus by varying r', (15) is replaced by the diffusion equation

(20) $$v_t = \frac{\varepsilon}{2}\,\partial_x^2(x^2(1- \frac{x}{K})^2 v) - \partial_x(\bar{r}x(1 - \frac{x}{K})v)\ .$$

Exercise: Apply Wright's formula to (20).

We can also vary K instead of r'. Let Q = 1/K. Then

(21) $$dX = r'X(1-QX)dt\ .$$

Let

(22) $$Q = \bar{Q} + \frac{dS}{dt}\ ,$$

where the last term represents the random component, as before. Then

(23) $$dX = r'X(1-\bar{Q}X)dt - r'X^2dS\ .$$

Exercise: Write down the corresponding diffusion equation and apply Wright's formula.

R. M. May (1973) derives a diffusion equation by a different method, which cheats a bit. In this procedure we rescale time,

$$\tau = \frac{r'}{K} t ,$$

and the logistic equation becomes

$$\frac{dx}{d\tau} = x(K-x) .$$

Now let

$$K = \bar{K} + \frac{dW}{d\tau} .$$

Then

$$dX = X(\bar{K}-X)d\tau + XdS , \tag{24}$$

$$E(dX|X = x) = x(\bar{K}-x)d\tau , \tag{25}$$

$$\sigma^2_{dX} = \varepsilon x^2 d\tau . \tag{26}$$

The density for $X$ satisfies the diffusion equation

$$\frac{\partial v}{\partial \tau} = \frac{\varepsilon}{2} \partial^2_x(x^2 v) - \partial_x(x(\bar{K}-x)v) . \tag{27}$$

Let $w = x^2 v$. Then when the flux is zero, we shall have

$$\frac{\varepsilon}{2} \frac{\partial w}{\partial x} - \frac{\bar{K}-x}{x} w = 0 ,$$

or

$$\log w = \frac{2}{\varepsilon} [\bar{K} \log \frac{x}{\bar{K}} + \bar{K} - x] + \text{const.}$$

Then

$$v = \frac{c}{x^2} \exp[\frac{2}{\varepsilon} (\bar{K} - x + \bar{K} \log \frac{x}{\bar{K}})] , \tag{28}$$

or

$$(29) \qquad v = \frac{c}{x^2}\left(\frac{x}{\bar{K}}\right)^{2\frac{\bar{K}}{\varepsilon}} e^{\frac{2}{\varepsilon}(\bar{K}-x)} \quad .$$

This solution may be singular at x = 0. For small x, it follows from (29) that

$$(30) \qquad v \sim c' x^{2\bar{K}/\varepsilon - 2} \; .$$

Thus v is integrable only if $\varepsilon < 2\bar{K}$. Near $x = \bar{K}$, $\log w \sim -\frac{1}{\varepsilon}(x-\bar{K})^2$, and for small $\varepsilon$, v is nearly Gaussian as in Fig. 1.

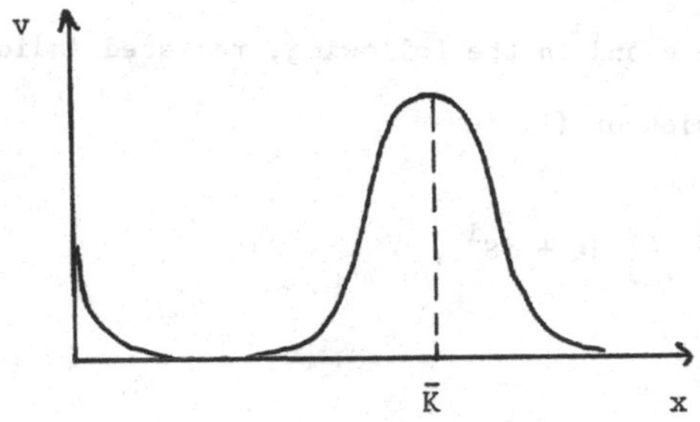


Figure 1

For larger $\varepsilon$, the peak at x = 0 becomes more prominent, as in Fig. 2.

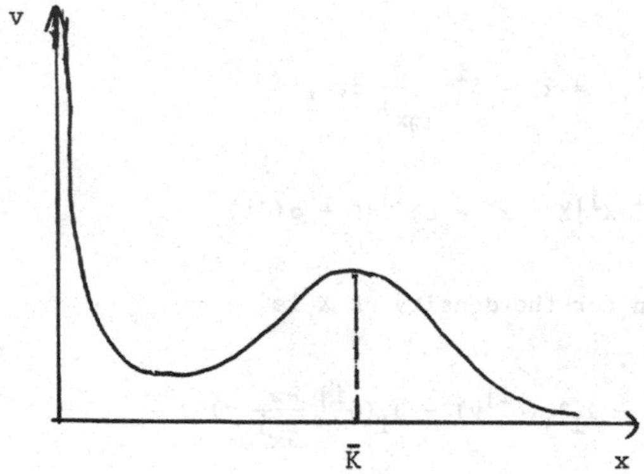


Figure 2

This situation is quite similar to what we saw before in the genetic load model. Then when $\varepsilon$ is larger and the density is not integrable, we might say that extinction is certain. However in the present case the basic equation (27) seems not to be valid at $x = 0$, and so that interpretation is questionable.

## 2. Several Species—Gradient Fields

We now consider a multidimensional case. X now is a vector, representing n species: $X = (X^1,\dots,X^n)$. For the deterministic case, we assume the special form

$$\frac{dX^i}{dt} = \delta^{ij}\frac{\partial W}{\partial x^j} \tag{1}$$

i.e. a gradient field. Here and in the following, repeated indices are summed from 1 to n. A stochastic version of (1) is

$$dX^i = \delta^{ij}\frac{\partial W}{\partial x^j}dt + dS^i , \tag{2}$$

where

$$E(dS^i|X = x) = 0 , \tag{3}$$

$$E(dS^i dS^j|X = x) = \varepsilon\delta^{ij}dt , \tag{4}$$

i.e. the variables $dS^i$ are uncorrelated. It follows that

$$E(dX^i|X = x) = \delta^{ij}\frac{\partial W}{\partial x^j}dt , \tag{5}$$

$$E(dX^i dX^j|X = x) = \varepsilon\delta^{ij}dt + o(dt) . \tag{6}$$

The diffusion equation for the density of X is

$$v_t = \frac{\varepsilon}{2}\partial_i\partial_j(\delta^{ij}v) - \partial_i(\delta^{ij}\frac{\partial w}{\partial x^j}v) . \tag{7}$$

When $v_t = 0$, this has the special solution

$$v = ce^{\frac{2}{\varepsilon}W} \quad . \tag{8}$$

This special form (7) of the diffusion equation was chosen since it has such a simple solution. This density has peaks near the relative maxima of W. From the deterministic equation (1)

$$\frac{dW}{dt} = \frac{\partial W}{\partial x^i}\frac{dx^i}{dt} \quad . \tag{9}$$

Therefore, neglecting random effects, the system wants to climb to the relative maxima of W (W increases except where $|\nabla W| = 0$). So, the relative maxima are stable equilibria. This behavior is reflected in the stochastic solution (8). For small $\varepsilon$, then the projection of v on one dimension is a curve such as

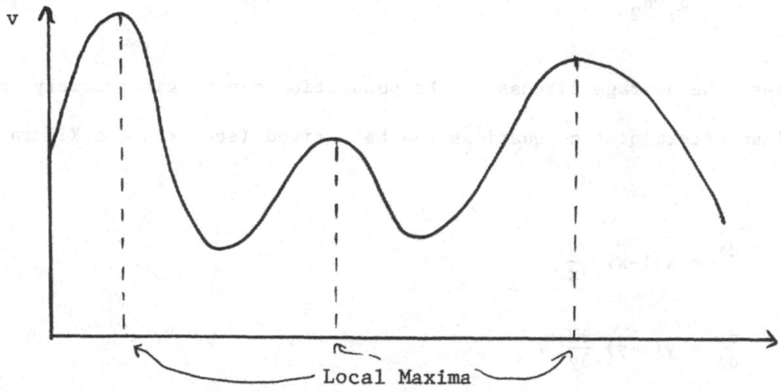


The smaller $\varepsilon$ is, the sharper the peaks and the larger the differences between heights. But, for larger $\varepsilon$ the reverse is true, and the chance of moving from one peak to another increases. This is an important difference between the deterministic and stochastic models. The following example illustrates a similar phenomenon in population genetics.

## A Genetics Problem

Consider 2 loci on two different chromosomes, with alleles $A_1$, $A_2$ and $B_1$, $B_2$ respectively.

Let

$$a_1 = \text{frequency of } A_1 ,$$

$$a_2 = \text{frequency of } A_2 ,$$

$$b_1 = \text{frequency of } B_1 ,$$

$$b_2 = \text{frequency of } B_2 ,$$

$$x = \frac{a_1}{a_1+a_2} ,$$

$$y = \frac{b_1}{b_1+b_2} .$$

In certain cases, the average fitness of the population can be given solely in terms of x and y. Then deterministic equations can be derived (see Crow and Kimura (1970), p. 181).

(10) $$\frac{dx}{dt} = x(1-x) \frac{\partial W}{\partial x}$$

(11) $$\frac{dy}{dt} = y(1-y) \frac{\partial W}{\partial y} .$$

If N is the population size, we have the diffusion equation

(12) $$v_t = \frac{1}{4N} \partial_x^2(x(1-x)v) + \frac{1}{4N} \partial_y^2(y(1-y)v) - \partial_x(x(1-x) \frac{\partial W}{\partial x}) - \partial_y(y(1-y) \frac{\partial W}{\partial y}).$$

The solution to this,is, by analogy with (8),

(13) $$v = \frac{c}{x(1-x)y(1-y)} e^{4NW} .$$

This expression is not integrable as it stands. To remedy this, we must add mutations, to make v integrable. In general, there are several relative maxima of W.

In analogy with the previous case, we expect v to be large near the relative maxima of W, or near the boundary (where one or more alleles is near fixation). Deterministically, the system will move to a relative maximum, and might never reach an absolute maximum of the fitness. However, if N is small (or if a subpopulation is isolated) then stochastic effects make it possible for the system to reach a higher maximum point. The importance of this effect has been emphasized by Sewall Wright (1964).

## 3. Ray Method for General Systems

The results of the previous section can be derived for a general system, which is not necessarily of the gradient type. This is an application of the Hamilton-Jacobi theory and the "ray method" which was developed for problems in wave propagation. We wish to solve equations of the form

(1) $$\frac{\varepsilon}{2}\,\partial_i\partial_j(a^{ij}(x)v) - \partial_i(b^i(x)v) = 0 \; .$$

We try a solution of the form

(2) $$v = e^{-\frac{1}{\varepsilon}\phi} z \; ,$$

which was suggested by the results of section 2. Then

(3) $$\begin{cases} \partial_j(a^{ij}v) = -\dfrac{a^{ij}}{\varepsilon}\,\phi_j v + O(1) \; , \\[2ex] \partial_i\partial_j(a^{ij}v) = \dfrac{a^{ij}}{\varepsilon^2}\,\phi_i\phi_j v + O(\tfrac{1}{\varepsilon}) \; , \\[2ex] \partial_i(b^i v) = -\dfrac{1}{\varepsilon}\,\phi_i b^i v + O(1) \; . \end{cases}$$

Ignoring the remainder terms, (1) and (3) imply the first order partial differential equation for $\phi$:

(4) $$\frac{1}{2}\,a^{ij}\phi_i\phi_j + b^i\phi_i = 0 \; .$$

We are thus naturally led to define the Hamiltonian

$$H(x,p) = \frac{1}{2} a^{ij} p_i p_j + b^i p_i \ , \tag{5}$$

with the corresponding system of Hamilton's equations

$$\begin{cases} \dfrac{dx^i}{dt} = \dfrac{\partial H}{\partial p_i} = a^{ij} p_j + b^i \ , \\ \dfrac{dp_i}{dt} = - \dfrac{\partial H}{\partial x^i} \ . \end{cases} \tag{6}$$

The rays are defined as solutions of (6). The Hamiltonian is conserved along rays, since

$$\frac{dH}{dt} = \frac{\partial H}{\partial p_i} \frac{dp_i}{dt} + \frac{\partial H}{\partial x^i} \frac{dx^i}{dt} = - \frac{\partial H}{\partial p_i} \frac{\partial H}{\partial x^i} + \frac{\partial H}{\partial x^i} \frac{\partial H}{\partial p_i} = 0 \ . \tag{7}$$

We shall choose initial values for x and p to ensure that $H \equiv 0$ along the rays. The function $\phi$ is defined on the rays by

$$\frac{d\phi}{dt} = p_i \frac{dx^i}{dt} \ , \tag{8}$$

which will actually imply that

$$p_i = \frac{\partial \phi}{\partial x^i} \ .$$

Now consider a surface S in the x-space on which $\phi = 0$. Then p must be normal to S. By setting $H \equiv 0$ on S (see (5)), we obtain an additional condition on p, which determines the magnitude of p.

Now p, x are known on S. Their values on S can be used as initial data for the system (6). If (8) is also integrated, then it can be shown that the solution gives parametric representation of the solution $\phi$: If S is parameterized by $\theta_1, \ldots, \theta_{n-1}$, then the solution of (6), (8), has the form

$$(9)\qquad \begin{cases} x(t,\theta_1\ldots,\theta_{n-1}) , \\ p(t,\theta_1\ldots,\theta_{n-1}) , \\ \theta(t,\theta_1,\ldots,\theta_{n-1}) . \end{cases}$$

If $t,\theta_1\ldots,\theta_{n-1}$ are eliminated from (9), the resulting $\phi(x)$ satisfies (4).

Reference: Courant (1962), Volume II, Chapter II.

<u>Point-Source Problem</u>

Now consider the special case where S shrinks to a single point, $x_0$.

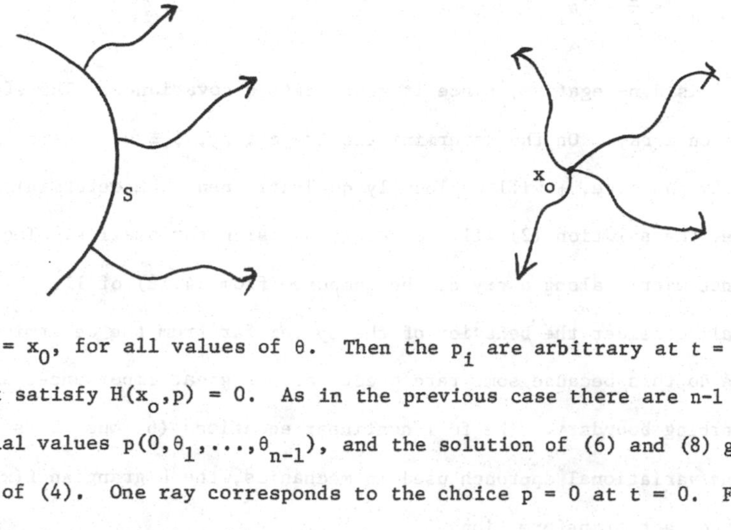


Let $x(0) = x_0$, for all values of $\theta$. Then the $p_i$ are arbitrary at $t = 0$, except that they must satisfy $H(x_0,p) = 0$. As in the previous case there are n-1 parameters for the initial values $p(0,\theta_1,\ldots,\theta_{n-1})$, and the solution of (6) and (8) gives a solution of (4). One ray corresponds to the choice $p = 0$ at $t = 0$. From the second set of equations in (6),

$$(10)\qquad \frac{dp_i}{dt} = -\frac{\partial H}{\partial x^i}$$

$$= -\frac{\partial a^{jk}}{\partial x^i} p_j p_k + \frac{\partial b^j}{\partial x^i} p_j .$$

It follows from the form of (10) that $p(t) \equiv 0$ is a solution of (10). The first set of equations in (6) then assumes the form

$$(11)\qquad \frac{dx^i}{dt} = \frac{\partial H}{\partial p_i} = a^{ij}p_j + b^i = b^i ,$$

Therefore, the deterministic trajectory corresponds to $p = 0$. Moreover, $\phi$ is quadratic in x near the deterministic trajectory. To see this, we note that from (6) and (8),

$$\frac{d\phi}{dt} = p_i \frac{dx^i}{dt} = p_i(a^{ij}p_j + b^i) , \tag{12}$$

or

$$\frac{d\phi}{dt} = p_i(\frac{1}{2} a^{ij}p_j + \frac{1}{2} a^{ij}p_j + b^i) . \tag{13}$$

Since $H \equiv 0$, (13) implies that

$$\frac{d\phi}{dt} = \frac{1}{2} a^{ij}p_i p_j \quad . \tag{14}$$

The matrix $(a^{ij})$ is non-negative, since it represents a covariance. Therefore $\phi$ cannot decrease on a ray. On the deterministic trajectory, $\phi \equiv 0$. Therefore if $(a^{ij})$ is strictly positive, $\phi$ will be locally quadratic near the deterministic trajectory, i.e. the solution (2) will be nearly Gaussian for small $\varepsilon$. The variation of the covariance matrix along a ray can be computed from (4.19) of II.

Now we shall consider the behavior of the system far from the deterministic trajectory. We do this because some rare events are of great importance, as in the case of an absorbing boundary. The full nonlinear equations (6) must be solved. According to the variational approach used in mechanics, the Lagrangian $L(x,\dot{x})$ is defined by the contact transformation

$$L(x,\dot{x}) + H(x,p) = p_i\dot{x}^i \tag{15}$$

Solving for p in terms of $\dot{x}$ and substituting, (6) implies that

$$\dot{x}^i - b^i = a^{ij}p_j , \tag{16}$$

and hence

$$p_i = a_{ij}(\dot{x}^j - b^j) , \tag{17}$$

where

$$(a^{ij})^{-1} = (a_{ij}) .$$

After substituting (5), (15) becomes

$$\text{(18)} \qquad L(x,\dot{x}) = p_i(\dot{x}^i - b^i) - \frac{1}{2} a^{ij}p_ip_j$$

$$= p_i a^{ij} p_j - \frac{1}{2} a^{ij} p_i p_j ,$$

$$= \frac{1}{2} p_i a^{ij} p_j , \quad \text{a quadratic form in } p .$$

Finally, in view of (17),

$$\text{(19)} \qquad L(x,x) = \frac{1}{2} a_{ij}(\dot{x}^i - b^i)(\dot{x}^j - b^j) .$$

According to Hamilton's principle (Courant (1962), Chapter II), or the principle of least action,

$$\text{(20)} \qquad \phi(x_1) = \min_{\substack{u(o) = x_o \\ u(T) = x_1}} \int_0^T L(u,\dot{u})dt .$$

Here the minimum is taken over all u which are differentiable and satisfy the boundary conditions at $t = 0$ and $t = T$. Moreover the u which minimizes the integral satisfies Hamilton's equations. This is an immediate consequnce of the Euler-Lagrange equations of the calculus of variations. These equations are derived as follows:

Let $u(t,\alpha) = x(t) + \alpha\delta u(t)$, where $x(t)$ is the minimizing trajectory. This is a one-parameter family of curves, with $u(t,0) = x(t)$. We use the first variation. Set

$$\frac{d}{d\alpha} \int_0^T L(u,\dot{u})dt \Big|_{\alpha=o} = 0 .$$

We have

$$L(u,\dot{u}) = L(x,\dot{x}) + \alpha \frac{\partial L}{\partial x^i} \delta u^i + \alpha \frac{\partial L}{\partial \dot{x}^i} \delta \dot{u}^i + \dots .$$

So, the first variation is

(21) $$\int_0^T \left(\frac{\partial L}{\partial x^i}\,\delta u^i + \frac{\partial L}{\partial \dot{x}^i}\,\delta \dot{u}^i\right)dt = 0 \ ,$$

where

$$\frac{\partial L}{\partial x^i}\,\delta \dot{u}^i = \frac{\partial L}{\partial x^i}\,\frac{d}{dt}\,\delta u \ .$$

Integrating by parts, (21) implies

$$\int_0^T \left(\frac{\partial L}{\partial x^i}\,\delta u^i - \frac{d}{dt}\left(\frac{L}{\partial \dot{x}^i}\right)\delta u^i\right)dt = 0 \ ,$$

or

(22) $$\int_0^T \left(\frac{\partial L}{\partial \dot{x}^i} - \frac{d}{dt}\,\frac{\partial L}{\partial \dot{x}^i}\right)\delta u^i dt = 0 \ .$$

The fundamental lemma of the calculus of variations now tells us that, since $\delta u^i$ is arbitrary, the integrand must vanish, and we obtain

(23) $$\frac{\partial L}{\partial x^i} = \frac{d}{dt}\left(\frac{\partial L}{\partial \dot{x}^i}\right) \ , \quad i = 1,\ldots,n \ ,$$

which are the Euler-Lagrange equations.

Now from (19) and (17),

(24) $$\frac{\partial L}{\partial \dot{x}^i} = a_{ij}(\dot{x}^j - b^j) = p_i$$

By differentiating (15), we obtain

(25) $$\frac{\partial L}{\partial x^i} + \frac{\partial H}{\partial x^i} = 0 \ .$$

Therefore (23) is equivalent to (6).

The above development is the well-known principle of least action in classical mechanics. The application to dynamical systems perturbed by noise was noticed by Ventsel and Freidlin (1970). We now consider their interpretation. The quantity L in our case is related to the likelihood that a given trajectory will be followed. Recall that

$$E(dx^i | X = x) = b^i(x)dt \ , \tag{26}$$

$$E(dx^i dx^j | X = x) = \varepsilon a^{ij}(x)dt + O(dt^2) \ . \tag{27}$$

Let dX be Gaussian:

$$\text{Prob}[\xi^i \le dX^i \le \xi^i + d\xi^i \ , \ i = 1,\ldots,n] \tag{28}$$

$$= K \exp[-\frac{a_{ij}}{\varepsilon dt}(\xi^i - b^i dt)(\xi^j - b^j dt)]d\xi^1 \ldots d\xi^n \ .$$

where K is a normalizing factor.

Let u(t) be an arbitrary trajectory. Then $du = \dot{u}dt$, and (28) becomes

$$\text{Prob}[dX \sim du] = K' \exp[-\frac{a_{ij}}{2\varepsilon dt}(\dot{u}^i dt - b^i dt)(\dot{u}^j dt - b^j dt)]$$

$$= K' \exp[-\frac{a_{ij}}{2\varepsilon}(\dot{u}^i - b^i)(\dot{u}^j - b^j)dt] \ ,$$

and hence

$$\text{Prob}[dX \sim du] = K' \exp[-\frac{1}{\varepsilon} L dt] \ . \tag{29}$$

If we ignore normalizing factors, the likelihood that a path X is close to u is given by

$$e^{-\frac{1}{\varepsilon}\int L dt} \ .$$

Thus the path that minimizes $\int L dt$ maximizes the likelihood.

As an application of these ideas, consider a stable deterministic equilibrium point $x_o$ in a domain D in which a system of n species can persist. The boundary points $x_1$ of the domain are points where one or more species become extinct, and hence the boundaries are absorbing. Considering all such points $x_1$, we draw rays from $x_0$ to $x_1$, and find

$$\phi(x_1) = \min \int_{x_0}^{x_1} L dt .$$

Then

$$\text{Prob}[X \text{ reaches } x_1] \sim e^{-\frac{1}{\varepsilon}\phi(x_1)} .$$

Picking $x_1 = x^*$ to make $\phi$ a minimum, we see that the likelihood of absorption maximum at $x^*$. Further details and interpretations are given in D. Ludwig (1975).

# B I B L I O G R A P H Y

Bailey, N. T. J. (1957) The Mathematical Theory of Epidemics. London: Griffin.

Bharoucha-Reid, A. T. (1960) Elements of the Theory of Markov Processes and Their Applications. New York: McGraw-Hill.

Courant, R. (1936) Differential and Integral Calculus, Vol. II. New York: Interscience (Wiley).

__________. (1962) Methods of Mathematical Physics, Vol. II. New York: Interscience (Wiley).

Crow, J. F. and M. Kimura. (1970) An Introduction to Population Genetics Theory. New York: Harper and Row.

Feller, W. (1951) Diffusion Processes in Genetics. Proc. 2nd Berkeley Symp. Math. Stat., pp. 227-246. Berkeley: Univ. of California Press.

__________. (1966) An Introduction to Probability Theory and Its Applications, Vol. II. New York: Interscience (Wiley).

Harris, T. E. (1963) The Theory of Branching Processes. Berlin: Springer.

Hoppensteadt, F. (1975) Lectures on Deterministic Population Theory. (To appear).

Karlin, S. (1966) A First Course in Stochastic Processes. New York: Academic Press.

Kendall, D. G. (1956) Deterministic and Stochastic Epidemics in Closed Populations. Proc. 3rd Berkeley Symp. Math. Stat. Prob., Vol. IV, pp. 149-165. Berkeley: Univ. of California Press.

Ludwig. D. (1974a) Final Size Distributions for Epidemics. To appear in Math. Biosciences.

__________. (1974b) Qualitative Behavior of Stochastic Epidemics. To appear in Math. Biosciences.

__________. (1975) Persistence of Dynamical Systems Under Random Perturbations. Submitted to SIAM Review.

MacArthur, R. H. and E. O. Wilson. (1967) Theory of Island Biogeography. Princeton: Princeton University Press.

May, R. M. (1973) Stability and Complexity of Model Ecosystems. Princeton: Princeton Univ. Press.

McKendrick, H. G. (1914) Studies on the Theory of Continuous Probabilities, with Special Reference to its Bearing on Natural Phenomena of a Progressive Nature. Proc. Lond. Math. Soc., Ser. II, 13, pp. 401-416.

Moran, P. A. P. (1962) The Statistical Processes of Evolutionary Theory. Oxford: Clarendon Press.

Murray, J. (1972) Genetic Diversity and Natural Selection. New York: Hafner.

Pielou, E. C. (1969) An Introduction to Mathematical Ecology. New York: Wiley:

Simberloff, D. S. and E. O. Wilson. (1970) Experimental Zoogeography of Islands. A Two Year Record of Colonization. Ecology, 51, pp. 934-937.

Ventsel, A. D. and M. I. Friedlin. (1970) On Small Random Perturbations of Dynamical Systems. Uspekhi Mat. Nauk, 25, pp. 3-55.

Voronka, R. and J. B. Keller. (1975) Asymptotic Analysis of Stochastic Models in Population Genetics. Submitted to Theor. Pop. Biol.

Waltman, P. (1974) Deterministic Threshold Models in the Thoery of Epidemics. Lecture Notes in Biomathematics. Berlin: Springer.

Watson, H. W. and F. Galton. (1864) On the Probability of Extinction of Families. J. Roy. Anthrop. Inst., 4, pp. 138-144.

Watterson, G. A. (1962) Some Theoretical Aspects of Diffusion Theory in Population Genetics. Ann. Math. Stat., 33, pp. 939-957.

Whittle, P. (1955) The Outcome of a Stochastic Epidemic. A note on Bailey's paper. Biometrika, 42, p. 116.

Wright, S. (1964) Stochastic Processes in Evolution, in Stochastic Models in Medicine and Biology, ed. J. Gurland. Madison: Univ. of Wisconsin Pres..

Yule, G. U. (1964) A Mathematical Theory of Evolution Based Upon the Conclusions of Dr. J. C. Willis, F. R. S. Phil. Trans. Roy. Soc. Lond., Ser. B, 213, pp. 21-87.

# INDEX